T0146119

The
Literature
of the
Life
Sciences

The Library and Information Science Series

This volume is the first of a series published by ISI Press®. The Library and Information Science Series provides timely and practical information to help librarians and information scientists.

Books published in this series:

The Literature of the Life Sciences: Reading, Writing, Research
    by DAVID A. KRONICK

The Literature of the Life Sciences

Reading, Writing, Research

David A. Kronick

**Assisted by
Wendell D. Winters**

iSi PRESS®

Philadelphia

Published by

ISi PRESS ®A Subsidiary of the
Institute for Scientifc Information®
3501 Market Street, Philadelphia, PA 19104 U.S.A.

© 1985 ISI Press

Library of Congress Cataloging in Publication Data

Kronick, David A., 1917-
   The literature of the life sciences.

   (The Library and information science series)
   Bibliography: p.
   Includes index.
   1. Life sciences literature.   2. Information
      storage and retrieval systems—Life sciences.
      I. Winters, Wendell D.   II. Title.   III. Series.
      QH303.6.K76   1985   574'.07   85-4283
      ISBN 0-89495-045-2

Printed in the United States of America

90 98 88 87 86 85   8 7 6 5 4 3 2 1

For
My dear wife and loving companion
Marilyn

Contents

Preface

The life sciences cover a vast spectrum of subjects and ideas and their relationships, including not only such directly related basic sciences as botany, parasitology, and zoology, but also the wide range of plant and animal species against which and through which the ideas derived from these disciplines can be exemplified and applied in agriculture, the health professions, and industry. Each of the disciplines, professions, and occupations associated with the life sciences has developed a vast literature of its own, and special manuals and guides have been produced for many of them. These literatures also have many characteristics and principles in common. These principles provide the focus for the discussions in this book. I have not tried to produce a comprehensive guide or an inventory of resources in any of the separate fields, or even to the life sciences in general, but rather to provide a general background of observations and useful information for the practitioner, investigator, and student in the life sciences, so that they will be able to read, write, and research the literature in a more perceptive and efficient manner.

As a result of the concern in recent decades about our ability to manage the expanding volume of scientific information, a large body of literature has grown up dealing with the problems of producing and using this information. Some of this literature has come from those who write scientific papers themselves, prepare them for distribution, or organize them for use. These principles have also been the focus of attention for a new discipline called "information science," which has been defined as a discipline "concerned with the generation, collection, organization, interpretation, storage, retrieval, dissemination, transformation and use of information, with particular emphasis on the application of modern technologies in this area" (195:5). These functions are also areas of concern for us in the discussions that follow. Although use of information comes at the end of the list cited above, it provides the rationale for the entire process and is the central focus for this book.

Science is, of course, a social as well as an intellectual activity, and many sociologists in recent years have turned their attention to the social milieu in which scientific workers operate and in which scientific ideas are generated and used. Many of the insights provided through these studies are also useful to the investigator and practitioner in the life sciences, making them more aware of how social factors influence the way they produce and use the literature.

I have tried to draw from this wide literature as well as from my reading and experiences in other areas to try to create a perspective to aid in understanding how the literature is produced, and how most effectively to use it. Throughout I have tried to maintain a historical point of view, both through examples from the literature and by reviewing some of the antecedents of our current scientific information systems, because I think it is important in helping us understand our current problems.

Writing and reading are two reciprocal processes as they relate to the scientific literature as well as to literature in general. Knowing how to read a scientific article, for example, being aware of the statistical and methodological criteria by which to evaluate it, can contribute immeasurably to knowing how to write one. While a scientific paper may not share many characteristics with a poem or a novella, it has its own stylistic requirements and characteristics which contribute toward its being recognized as good or bad scientific writing.

Investigators, clinicians, and practitioners of all kinds who draw their information from the life sciences need to gain some awareness of how knowledge is transmitted and how it is organized for retrieval and use, so that they will be able to use centralized and institutionally based information systems effectively. In the process they will inevitably collect information sources of their own in the form of reprints of scientific papers, notes, and other documents. These should be organized for effective use by developing personal information systems. These systems should be congenial and comfortable to the working habits, styles, and needs of the individual. They can also be more effective in terms of time and energy invested if they can be consistent with and utilize the products of shared systems. Too many personal information systems have collapsed because they have proved too time consuming and cumbersome. In the chapter on personal information systems, I have tried to draw from the reported experiences with personal information files, as well as to suggest some of the opportunities offered by the newer technologies.

One of the newer technologies, or rather newer methods, of organizing, accessing, and evaluating the literature is citation indexing. This technique provides us with a method for qualitative filtering of the literature. It cuts across the entire subject range of recorded information, including the arts and humanities as well as the sciences. Citation indexing is particularly valuable in the sciences, however, because of the cumulative nature of the subject matter and should be a part of the armamentarium of every serious student of the life sciences.

My intention here is not to provide definitive guides to the literatures of chemistry, biology, and medicine. These subjects are all central to and provide the basic pools of information for the study of the life sciences. Many comprehensive manuals and guides exist for all these subjects and are constantly being issued or reissued. Instead, I have tried to deal with the basic issues concerning the information systems in each of these disciplines, including some history of their development, their special terminology and vocabulary problems, and the kinds of special information sources they use. I have also included a description of some of the primary information resources in each of these disciplines, in terms of the kinds of issues raised throughout the book.

The computer and other machine aspects of information production, storage, and accessibility are not dealt with separately, because they are beginning to pervade every aspect of the literature of the life sciences. They promise to have even greater impact in the coming years. Although it is always hazardous to predict the future, it is becoming clear that the new technologies in information handling and communication may have the capability of radically changing the way we store, transmit, and use information. It also seems likely that they may only provide us with different and improved tools to enhance or modify the essentially intellectual task of creating new knowledge and putting it to work in solving our scientific and social problems.

In summary, I hope that all serious students of the life sciences, including undergraduate and graduate students in the life sciences, dental, medical, nursing, and other health professionals, investigators in all related fields, and practitioners in all related professions, will find some useful information in the following chapters to help them toward achieving these objectives:

1. To be able to select the appropriate kind of literature source to solve a particular information problem.
2. To be able to formulate the problem in researchable terms which are congenial to the information system used, and which can be articulated in an understandable way to others.
3. To be familiar with some of the principal guides to the retrospective and current literature.
4. To have an understanding of the structure of indexes and other kinds of keys to the literature and the principles on which they are based.
5. To gain some insights into the way the literature of the life sciences is produced, stored, and utilized.
6. To have some familiarity with the use of computer-based systems in literature searching.
7. To gain some historical perspective on how the literature has developed and how it has been utilized.

References are all given in a single numbered list at the end of the book because some of the references are cited in more than one chapter. They are arranged in alphabetical order by author or, in the case of anonymous works, by title, and they are cited in the text by number. When the citation is to a particular page, the page number occurs after the colon following the reference number.

I owe special thanks to my friend Wendell D. Winters, who provided encouragement and advice. His experience as a productive and successful bench scientist meant a great deal in sustaining me in carrying this book through to its conclusion.

As usual, my secretary Theresa Jimenez contributed beyond the call of duty in using her extraordinary skills and talents to produce the manuscript.

I have drawn freely and unashamedly from the informative essays with which Eugene Garfield has for years been prefacing the issues of *Current Contents,* which are collected under the title *Essays of an Information Scientist* (152). His humanistic approach to the literature of science and to science in general as a high calling provided much of the inspiration for this book.

Chapter 1

The Literature of the Life Sciences: A Perspective

Roles and Functions

The literature of the life sciences plays many roles and fulfills many functions for investigators and practitioners in the basic and applied sciences. It has been compared to the human nervous system (80:16), which is constantly receiving and transmitting signals and stimuli to and from all parts of the body. There are, however, important differences since the literature encompasses only a part of the total information being transmitted. Information is also transmitted by other random and organized sources such as chance and arranged meetings with colleagues. Information can be transmitted across time from one generation to another, across space between widely dispersed geographic areas, and from one interest group to another.

A primary role of the literature is to record and transmit discoveries or ideas that advance the state of knowledge. Sometimes parts of these communications remain latent and inert in the literature. This can be beneficial when portions of the literature are irrelevant, nonsensical, or soporific. When these information spores, like dormant bacteria, find a fertile medium in the minds of receptive investigators, they can stimulate research. A classical example is Gregor Mendel's article on plant breeding which was ignored for almost 40 years, from the time it was published in 1866 until 1900 when it began to influence a new era in genetic research. The failure of Mendel's results to influence thinking about genetic problems was no accident. Although published in an obscure journal, Mendel's article was circulated to 120 learned societies and universities. No place, however, for Mendel's work could be found in the general framework of biological theory at the time. Not until DeVries, Correns, and Tschermak all thought that they had independently reached the same conclusion around 1900 did they become aware of Mendel's work (176).

Another important role of the literature is to help to solve problems in research or practice. The ability to solve a problem depends to a large extent on possessing the information necessary to solve it, either by means of prior

1

knowledge or by being able to gain access to existing knowledge. Some individuals may be fortunate or diligent enough to have sufficient information stored in their own memories to solve problems as they occur in research or practice. However, most of us must access external sources. The literature has been described as a vast external memory in which all human experiences (observations and experiments) and ideas have been recorded since the beginning of permanent records. Indeed, there are some intimations that external memories may function in many ways similar to individual internal memories. Information scientists, psychologists, and neurophysiologists are far from bridging the gap between their various approaches to the problems of memory, although developments in computer technology and research in all three areas are filling in many of these gaps.

One advantage of external memories, the literature and the various devices and techniques that have been used to organize it and access it, is that they can be studied more easily than internal organic memories and, consequently, can be more easily understood and modified. Another important characteristic is that they are systems that are shared among many people. It is necessary, therefore, to be aware of what assumptions we share about the literature and to be knowledgeable about the conventions we have developed for producing and utilizing it.

Our ability to solve problems has a direct relationship to the amount of relevant information we have. However, too much prior knowledge can stifle creativity and invention. One illustration comes from outside the life sciences. The builder of the Gossamer Condor, the first aircraft to fly under human power, is quoted as saying:

> I've come to appreciate more and more the stifling influence of old thought, old ideas. Professionally qualified people, big teams with many credentials and plenty of resources, built planes for the Kremer Prize. But because they were experienced they built them the way planes had been built before. So their aircraft were complicated and heavy and failed. Their prior knowledge stifled innovation. (139:72)

There is something persuasive and appealing in the idea, but it ignores the fact that Paul MacCready, who won the Kremer Prize in 1977 for this feat, was also the head of a firm which build air drag reduction devices and conducted wind power studies and that he was a champion glider pilot who built model planes as a hobby. Hermann von Helmholtz (1821–1894), who started as a humble army surgeon, became progressively an outstanding physiologist and physicist and made fundamental contributions in physiological acoustics, physiological optics, and pure physics. He has been called "one of the last individuals in the history of science whose work embraced all the sciences." (440). Were his contributions due to the fact that he was not inhibited by the conventional wisdom in the fields he entered, or did they result from his ability to apply prior knowledge to the solution of new problems?

We must also be aware that scientific information is transmitted through other means than the literature. Information may be absorbed as part of a shared "belief system" in the anthropological sense. The literature as an external memory system shares other characteristics with personal memories in that it may also contain information, ideas, and beliefs that are socially determined rather than verified by experience or in the laboratory, and these may be false or misleading. Controversy continues about the extent to which the development of scientific ideas is influenced by social as well as intellectual factors. It is well recognized, nevertheless, that science can be studied as a form of social activity. Science has fads and fashions similar to those that can be seen in dress, art forms, and other social expressions, and it is subject to the same kind of peer pressures (447). There seem to be the same needs for conformity in science that there are in society, as can be seen by the bandwagon effects in areas of research funding and in the use of specific technologies.

Robert Merton, a pioneer in the application of sociological methods to science as an activity, has identified four basic social norms which are inherent in the scientific ethic:

1. *Universalism*—scientific ideas have validity, no matter what the geographic, racial, or religious background of their proponents, and these ideas are subject to criteria which derive only from science.
2. *Communism, or communality*—scientific ideas are a shared commodity that belongs to everyone.
3. *Disinterestedness*—scientific ideas are pursued not for the sake of personal gain but for the enrichment of public knowledge.
4. *Organized skepticism*—no ideas are sacrosanct but all are open to investigation and inquiry (309).

Merton points out that although these norms are generally agreed to they are not always observed in practice. In fact, it seems that, like some of the Ten Commandments, they may be honored as much in the breach as in the observance. The idea of universalism is challenged by many instances of secrecy and withholding information, a practice which is encouraged by the highly competitive nature of granting mechanisms, the industrial base of some research efforts, and the claims of national security. The notion of communality of scientific knowledge is denied by the many disputes about priority of discovery which occur in science and by the sensitivity demonstrated by authors whose ideas have not been given adequate recognition in the literature. A contradiction to the norm of skepticism can be found in the educational process in which students are often taught to answer rather than to ask questions. In fact, one of the polarities in science characterized by Kuhn as the "essential tension" is exemplified by the observation that "the successful scientist must simultaneously display the characteristics of the traditionalist and of the iconoclast" (264:277). The sociology of science shares some of its precepts with the sociology of knowledge: both make us aware that we tend to think and reason in ways that we are socially conditioned to do so, and that we tend to believe what we want or need to believe.

The literature serves functions other than the disinterested, communal sharing of information, since recognition and other rewards in science are achieved largely as a result of published work. It is an interesting paradox that "the more widely scientists make their intellectual property freely available to others, the more securely it becomes identified as their property" (307).

Finally, in some cases the literature of the life sciences can be read for aesthetic pleasure. Although in its modern form, it frequently exudes little more charm and excitement than a weather report, the history of science provides many examples of aesthetic and poetic responses to scientific experiences. Writing on his microscopical investigations in 1665, Robert Hooke exclaimed: "These pleasing and lovely colours have I also sometimes with pleasure observ'd even in Muscles and Tendons" (130:57). It is also hard to miss Malphigi's excitement when he quotes Homer in a letter to his friend after having seen the capillaries for the first time: "I see with my own eyes a certain great thing" (141:95). The Roman philosopher Lucretius (99–143) wrote his great comprehensive work *De Rerum Naturae,* which summarized the scientific thought of his time, in the form of a poem. The great eighteenth-century Swiss scientist Albrecht von Haller's (1708–1777) reputation as a poet is as great as his reputation as a physiologist. Claude Bernard (1813–1877) began his career with a desire to be a dramatist, but he revolutionized the study of experimental medicine instead. In his book *La Science Experimentale,* published in Paris in 1865 (the same year in which his celebrated *Introduction to the Study of Medicine* appeared), he said:

> Science does not contradict art. I cannot accept the opinion of those who think that scientific positivism must be fatal to inspiration. Quite to the contrary: the artist will find in science a more stable formulation, and the scientist will draw from art a more certain intuition. (450)

Many scientific writers have also been regarded as a fine literary stylists. It has been said of William Osler's (1849–1919) celebrated textbook *Principles and Practice of Medicine,* first published in 1892, that it "succeeded in making a scientific treatise literature" (209:4). Sir Peter Medawar, who is himself a fine stylist, has referred to D'Arcy Thompson as possessing "a beauty and clarity of writing that may never be surpassed" (300:35). One can pick a passage from D'Arcy Thompson's *On Growth and Form* almost at random as an example of beautiful simplicity:

> The eye and its retinal elements have ranges of magnitude and limitations of magnitude of their own. A big dog's eye is hardly bigger than a little dog's; a

squirrel's is much larger, proportionately, than an elephant's and a robin's is but little less than a pigeon's or a crow's." (435:53)

Models and Metaphors

Regarding the scientific literature as a form of external memory is one of many models and metaphors which have been applied to its development and organization. Derek Price lists among these metaphors: scientific subjects as geographic areas, each with its own boundaries and adjacencies, or as trees with roots and twigs, and science itself as a pyramid of brick-like discoveries (370:554).

The literature as external memory is one of the more powerful metaphors since it illustrates some of the ways we can regard our scientific information storage and retrieval systems and suggests that the processes of personal memory may be similar to those used in shared or external memory systems.

Other metaphors describe science as a structure made up by the additions of small segments to a puzzle from which a big picture emerges. The psychologist Herbert Simon reflects this point of view: "The product of a research program is a mosaic of particular results, for the requirements of journal publication force the scientist to exhibit each separate title, leaving the whole pattern to the imagination or whim of readers" (408:x). This concept is also reflected in what has been called the "Ortega Hypothesis," from Ortega y Gasset's contention that science advances by many small discoveries (344). In the same context the Nobel Laureate H. W. Florey, one of the discoverers of penicillin, is cited as saying:

> Science is rarely advanced by what is known in current jargon as a "break-through," rather does our increasing knowledge depend on the activities of thousands of our colleagues throughout the world who add small points to what will eventually become a splendid picture much in the same way the Pointillistes built up their extremely beautiful canvases. (80:217)

The Ortega Hypothesis is also reflected in an aphorism still incorrectly attributed to Newton, which Merton has engagingly pursued in his "Shandean Postscript" on the origins of the aphorism: "If I have seen further it is by standing on ye shoulders of Giants" (310). It is also probably true that a sufficient number of midgets standing on the shoulders of other midgets may attain the stature of a giant. All in all, this metaphor may be an ornament to the scientist's sense of modesty, but it does little more than underline the necessity for establishing and maintaining continuity in scientific thought and perhaps suggest the possibility that one scientist may be blessed with the opportunity to make a more significant contribution to its growth and transformation than another.

Another metaphor or analogy offered a few years ago compared the growth of knowledge with the metabolic processes. It is particularly apt if you speak of the "body" (corpus) of literature as being capable of carrying out such processes as assimilation, digestion, and decay, and having obsolescent or even waste products:

Information is one of the vital nutrients of research. Its digestion, absorption and assimilation by the scientific community may be likened to the metabolism of food by the human body. (342:15)

Paul Weiss has referred to this as a process of progressive synthesis, also in biological terms:

Organic growth is by assimilation, not accretion. Food items are not simply stuck on the body, but on the contrary, lose their identity and become anonymous and indistinguishably blended into the body's very own type of constituents by the processing chain of extracting, screening, sorting, fitting and recasting. . . . Accordingly, if knowledge grows like organisms, we ought to observe sound dietetics and avoid unhealthy overstuffing, and the symptoms of glutredundancy. (460)

This does not rule out *healthy* overstuffing, but it does make it clear that "literature" is not an equivalent of "knowledge."

A parasitologist working with an information scientist developed a model which compared the growth of the literature with an epidemic process in which scientific articles and the journals in which they are published serve as infective agents with readers as susceptible hosts. The model provides an interesting analogy to sociological studies in which rumors have been shown to be filtered and distorted through the perceptions of those who transmit them. Later, finding the infectious disease concept rather pejorative, since it tended to reinforce a view that the formal system of scientific information was in disorder if not actually in an unhealthy state, the authors modified the model to that of an "ecosystem of scientific communication." This described the flow of manuscripts and published papers between authors and journals, with libraries and societies as intermediaries, in ecological terms, but frankly does not seem nearly as picturesque or informative as the other model (185).

A recent model which has attracted significant attention is that of Thomas S. Kuhn (266). A great scientific revolution (usually referred to as "the scientific revolution") is said to have occurred in the seventeenth century. It involved a general change in modes and styles of thinking, a turning away from entrenched authoritarian thought, and an insistence on verifying facts through careful observation and experiment. Kuhn's thesis is that the development of scientific thought occurs not through any single revolution but through a succession of many revolutions which take place at various times in different areas of research. We tend to regard a scientific discovery, he says, as a unitary event and to attribute it to a single individual. Discovery frequently represents, however, the culmination of a series of events. Furthermore, scientific discoveries are of different kinds, those which could be predicted from existing theory and those which could not be. The latter are the ones that upset the currently prevailing model which is established when a general consensus emerges from a group of knowledgeable co-workers in a particular area of research. The consensus establishes a picture, a model, or, in Kuhn's word, a "paradigm" which describes the way in which investigators in the field understand the phenomena with which they work.

The consensus introduces a period which Kuhn calls "normal science" in which investigators in the field continue to work within the conceptual framework or paradigm. An anomaly occurs when someone, either working in that particular field of inquiry or outside it, introduces an observation or concept which threatens the consensus and which may lead to the restructuring of the model. When a new consensus is achieved, it is followed by another period of "normal science" (192). The physicist Niels Bohr reflects this point of view when he says that one must be aware that "all knowledge presents itself within a conceptual framework adapted to account for previous experience and that any such frame may prove too narrow to comprehend new experience" (46:67–8).

Variations of these ideas occur elsewhere in the literature. For instance, Cournand says there are three kinds of scientists: the first are investigators who work inside the boundaries of "normal science" and provide us with the initial level or type of discovery of which Kuhn speaks, the second are the verifiers and consolidators who write the reviews and compile the systems and textbooks, and the third are the revolutionaries who provide the "mutation in scientific thought" (92). These comments point to the tentative nature of all knowledge, including scientific knowledge. "There is no absolute knowledge," says Bronowski, "and those who claim it, whether they are scientists or dogmatists, open the door to tragedy" (63:60). Conant seems to concur when he implies that scientific ideas which have reached the state of consensus at which they are no longer the objects of investigation are no longer science but dogma (84:24).

Scope of the Literature

Knowledge, like reality, constitutes a unity; that is, to study any aspect of nature we need to isolate it from the complex relationships in which it exists. To understand our perceptions of reality, we need to compartmentalize, organize, and classify the phenomena which reflect them. We tend to forget sometimes that these efforts can only provide us with approximations of reality since classifications are merely arbitrary and pragmatic devices to help us organize our perceptions. Classifications are unidimensional, while knowledge is multidimensional. Concepts can exist in more than a single spectrum, axis, or relationship with other concepts. We must also be aware that the bodies of concepts we are trying to subsume under the literature of the life sciences are constantly in a state of change in both their content and their relationship to other bodies of concepts. Some may endure for a long time; others may be quickly displaced.

The life sciences constitute a part of that area of human thought and endeavor which we have come to call scientific. It is characterized by its basis of empirical and reproducible evidence, and also by the ways in which consensus is achieved. It is primarily the nature of the consensus and the methods by which it is achieved that distinguish scientific thought from other areas of thought.

Basic distinctions have been made between the sciences and the humanities. Fears have been expressed that these distinctions have created two separate worlds of thought and that inability to reconcile or relate them is detrimental to society. In science, hypotheses must be tested by controlled experiments in the laboratory, and pure speculation is discouraged; in other areas such as literature and philosophy, and in the humanities in general, intellectual activity proceeds by means of argument and debate rather than experiment. Although scientists dating from the seventeenth century, including Newton and Boyle, have tried to reconcile their religious and scientific views, they have tended to regard these areas essentially as separate universes of discourse, with their own vocabularies and methods of arriving at consensus. Humanists, it has been said, are not concerned as much with the accumulation of facts as with changes in perspective relating to existing sets of facts or perceptions which are then reviewed at different angles or in a different light (183:1634).

The difference between scientists and humanists, says Garfield (170:40), is that scientists look for new facts which can be reproduced through experiments. Each such fact (verified data) adds to a store of knowledge which the scientist draws on to generate other new facts which may supersede the old ones. In this sense it is a cumulative and evolutionary process. Scientific research, Garfield adds, echoing Kuhn's concept of the paradigm, is carried out in order to fill gaps in an existing model. For this reason, scientific communications tend to be short. Since scientists tend to build on previous work, they have developed a tradition of recognizing (referencing) the earlier work which influenced them. They also tend to be concerned primarily with the most recent work which reflects the results of the last stage of the evolutionary or cumulative process. The humanities literature, on the other hand, grows out of constant efforts to reevaluate past experience and thus tends to use and to cite a much older literature. Changes come as radical shifts in thought rather than as a process of orderly accumulation of new facts. There also exists less of a consensus; there is more room for alternative points of view on the music of Bach than on viral replication. The polar concepts are expressed by quotations from Horace, who said "There is a measure of all things," and from Protagoras, who said "Man is the measure of all things." These quotations occur in a recent book, *Toward a Metric of Science,* which has as a frontispiece an illustration from a 1536 book on geometry that defines a rod as the feet of 16 men ("short men and tall men") placed one after another (126).

Throughout the centuries, knowledge has been subdivided into increasingly smaller segments for study and analysis. The process itself goes back at least to the medieval period, with the creation of the trivium (grammar, rhetoric, and logic) and the quadrivium (arithmetic, music, geometry, and astronomy). Scholars and scientists have always dealt with the ever expanding areas of knowledge by a process of fragmentation and specialization which has been described as a philosophy of "divide and conquer." "It is only through this fragmentation process that their strength and pressure can be brought to bear upon a small enough facet of the universe to yield to their probing and understanding" (353:4).

Attempts have been made to place all these segments of knowledge, particularly the sciences, along a spectrum or continuum, which to some degree reflects the complexity of the phenomena (number of variables) with which it deals. The continuum also represents the degree to which each discipline provides a basis for understanding the phenomena in the next discipline, not only in terms of level of structure—atom, cell, tissue, organ system, organ, organism— but also in terms of the kind of phenomena they measure and the degree of accuracy with which the phenomena can be measured.

The place of a knowledge area (subject) on the spectrum has also been defined in terms of what is called "codification." Codification is represented, say Zuckerman and Merton (483:507), by the degree of "consolidation of empirical knowledge into succinct and interdependent theoretical formulations," that is, upon the agreement on general laws which apply to the phenomena observed. "In a scientific field where a large percentage of the papers submitted for publication are accepted, there is a higher level of codification, although rejection rates are certainly not the only measure of codification" (177:23). Codification tends to be higher in physics and chemistry than in botany or zoology, and again is related to the complexity of the materials involved (the greater number of variables which must be identified and controlled). Codification is also related, undoubtedly, to the ease or difficulty in achieving consensus.

There is some question about the importance of these formulations in dealing with problems in the research laboratory or in the real world. One frequently needs to draw from widely dispersed parts of the broad spectrum of knowledge in solving any problem, and many problems are attacked through cross-fertilization of ideas and multi-disciplinary approaches. The history of science is full of examples, such as that of Carleton Gadjusek, who solved the puzzling problem of the origins of Kuru through his combined knowledge of anthropology and virology (140:23).

In some ways knowledge progresses through the process of building bridges between disciplines, as Simon suggests when he talks about the different approaches to the study of human information processing which involve studies of the neural mechanisms by neurophysiologists as well as studies of information processing by psychologists. He compares these disciplines (areas of study) to nineteenth-century chemistry and physics which existed as independent bodies of theory until bridges began to be built between them.

At present, we have almost no knowledge of how to build a bridge from one to the other, although most of us have no doubt that such a bridge will be constructed by future generations. Meanwhile, the lack of a bridge need not deter us, and has not deterred us, from vastly broadening and deepening our understanding of human cognition in information processing terms. (408:xi)

The compartmentalization and classification of scientific endeavor also occur in the use of the terms "basic" and "applied" to characterize research efforts. On this subject Pasteur said: "Applied sciences do not exist; there are only applications of science" (354). The National Academy of Sciences study of the Life Sciences recognized:

> ... that there is no meaningful close definition of the terms ... and that these indications by our respondents reflect their motivations in addressing specific problems and not the character of their work. By this measure, one investigator studying sodium transport in human erythrocytes may classify it as "basic" research; another may consider the same study "clinical" only because human cells are employed for the purpose; and a third may view it as applied, since he hopes to develop a new drug. (318:263)

The apposition or dichotomy is also reflected in what is called "mission oriented" or "discipline oriented" literature services, the latter represented by such services as *Index Medicus,* a comprehensive index which attempts to cover all the clinical and related basic science literature, and the first by special indexes which were developed to serve the investigators of various categorical diseases (8:308).

The National Academy of Sciences also recognized that the conventional divisions of the field of biology were no longer an adequate basis from which to approach the study of the life sciences. Instead of the classical categories such as zoology, botany, and microbiology, they divided the field into such areas as molecular biology and biochemistry, developmental biology, etc., "based on the work of groups of like-minded scientists who have blended the approaches of several older disciplines in attacks on some specific subsets of biological problems" (318:276). Examples they gave of such subsets were photobiology, neurosciences, oncology, vascular physiology, and environmental health. They also pointed out the strong links between the life sciences and the physical sciences, "particularly in the use of physical instrumentation and in the pervasiveness of biochemical concepts and techniques. . . ." (318:iv). They found that the old academic department designations no longer served, if they ever did serve, to define the activities and research interests of the members of those departments. They collected 195 distinct departmental titles in their survey, and 300 additional titles used with subunits. They finally decided to recognize 100 subdisciplines under the life sciences for the purposes of the study, grouping them into twelve major fields. Although this categorization was useful for the purpose of the survey, we must recognize again that it may be as arbitrary and limiting as the categorization into the traditional disciplines and that it represents but a single cut across the body of literature that we have designated as the literature of the Life Sciences.

Coping with the Literature

In addition to the problems of defining its scope and subject relationships, two other factors complicate attempts to cope with the literature of the life sciences: its phenomenal growth and, perhaps as a corollary, its rapid obsolescence. I shall discuss these along with some other factors in Chapter 5, "Characteristics of the Literature." An even more difficult problem is how to maintain a high level of quality in the literature without inhibiting freedom of expression.

Everyone has heard expressions highly critical of "the literature," for example:

> On his first day in the laboratory, the observer was greeted with a maxim which was constantly repeated to him in one or another modified form throughout his time in the field. The truth of the matter is that 99.9% (90%) of the literature is meaningless (crap). (272:258)

This summary of several observations was made by a French anthropologist who went into a biological research laboratory with the same spirit of inquiry and curiosity he would use in exploring a community of South American aborigines. This same observer also described the intensity of the work, the difficulty of arriving at meaningful results, and the tenuous nature of most scientific information. At the end of one day's work in the laboratory, he commented:

> One or two statements have seen their credibility increase (or decrease) a few points, rather like the daily Dow-Jones Industrial Average. Perhaps most of today's experiments were bungled, or are leading their proponents up a blind alley. Perhaps a few ideas have become knotted together more tightly. (272:16–17)

Perhaps we can find some solace and instruction from history in knowing that scholars and scientists have always had these kinds of anxieties about the growth and quality of the literature. The unknown prophet in Ecclesiastes cried out "Of the making of books there is no end and much study is a weariness of the flesh" (123). Christian Gottfried Gruner (1744–1815), Professor of Medicine at the University of Jena for 40 years and himself an editor, echoed the complaint in somewhat different language: "The majority of physicians are surfeited to the point of illness by the reading of journals" (260:247).

One way the scholar–scientist throughout the ages has dealt with the problem of the exponential growth of the literature has been through the processes of specialization and delegation. Although there is a direct relationship between the growth of the literature and the growth of populations, there are obviously self-limiting factors in projections of trends in each. For instance, the growth projected for workers in the field of electronics in the 1960s meant that the entire work force would be involved in that industry by 1990 (28). Price's projection of the growth of scientific journals by a factor of ten every 50 years would mean that there would be a million by the year 2000 (372:93).

We have come a long way since 1592, when Francis Bacon could exclaim, "I have taken all knowledge to be my province" (52:65). The growth of specialization in scholarship can be exemplified by the third edition of the *Ency-*

clopedia Britannica in 1788, which for the first time was the work of a corps of specialists rather than one or two individuals (28:161). By the end of the eighteenth century the French scientist Antoine-Francois Fourcroy (1755–1809), said to be the father of biochemistry, wrote: "It is necessary to separate the three parts of the curative arts, pharmacy, chemistry, and medicine, for which the confusion can only bring about a triple mediocrity" (279:538).

In this environment the process of selecting the appropriate literature becomes even more critical. Literature which is irrelevant to the solution of a particular problem can be regarded to some extent in the same way a gardener defines a weed: something which is growing where it should not be growing. The scholar may begin by cultivating the attitude of the English poet writing at the time of John Milton on quite a different subject. He begins by asking:

> Shall I, wasting in despair,
> Die because a woman's fair?

and ends by avowing:

> For if she be not for me,
> What care I for whom she be. (469)

This may be overstating the case somewhat, for workers in the life sciences still need to maintain an awareness of the current state of the art in their primary areas of concern. They also need to access quickly and efficiently that segment of the literature which is relevant to their problem. To accomplish this, they must learn how the literature is currently produced, disseminated, and organized.

When the first Sputnik went up, a feeling was generated that Americans were falling behind the Russians scientifically. As a result, investigators from many disparate fields joined librarians and bibliographers whose traditional function has been to cope with the problems of information handling. The recruits came from fields like engineering, linguistics, computer sciences, and sociology, all of which had something relevant to add to the discussion. If the process of research and publication could be referred to as "the institutionalization of the production of scientific truth," it was argued, it could also be studied as a sociological problem (82:205). The new discipline which coalesced around the concept of managing information is called information science. As a result, we know a great deal more about the characteristics and the use of the literature. It has even introduced a new anxiety which Jesse Shera has called "keeping up with keeping up" (401).

To survive successfully in this new "information rich" society, every practitioner, teacher, student, and investigator in the life sciences must learn effective methods of gaining access to and using the literature which is relevant to their needs. It is with the task of sharing some of the knowledge of the characteristics and nature of this literature, its writing (production), reading (evaluation and selection), and research (retrieval strategies and techniques) that this volume is concerned.

Chapter 2

The Historical Background*

To understand any phenomenon, it is useful to know its antecedents. This is a principle which physicians apply when they take a history from their patients to establish a diagnosis. This approach may also be relevant to understanding the "body" or "bodies" of literature in the life sciences as the terms are used in such expressions as the "Hippocratic Corpus" in referring to the literary works of the legendary and historic Greek physician who, in another biological allusion, is called the "father of medicine." The expressions imply an organic nature to the literature and suggest that, like all organic forms, it has gone through a process of evolution which can be described historically. They also imply that, by studying the evolution of the medium, we can more easily understand the more complex forms we see today. In the development of scientific communication, as in all evolutionary processes, some of the species have survived and continue to function, although in some cases with changes attributable to changes in the social and scientific environment. A British medical historian made the argument for knowing the history of a phenomenon in another, yet similar, manner. He compared history to a Bach fugue, because history, he said, "possesses a unique power of showing how the complexities of the present have evolved from relatively simple roots. It displays the logic of this growth, much in the same way as Bach presents us with a simple theme before confounding us with the contrapuntal wizardry of his fugue" (242:258).

In the history of scientific communication and of communication in general, no communications technique ever seems to be completely obliterated. In the historical antecedents of our present communication systems, scientists and practitioners will easily recognize prototypes of media and communications practices which they use today. Oral transmission of information, our most ancient medium, is still widely used, especially since it has been enhanced and facilitated by electronic means and by the ease and speed of travel. In fact,

* Reprinted, with changes, from the *Bulletin of the New York Academy of Medicine* 60, no. 9 (November 1984):857-875.

science historians lament the lack of the written records available in the past in the form of correspondence and diaries, which have been supplanted in some cases by the telephone. As a result, our perceptions of how research is conducted and how discoveries are made may be distorted. Historians of contemporary science must depend on what they call oral histories, i.e., interviews with individuals who have been involved in particular scientific events.

Communications systems are the result of the interactions between the perceived needs of their users and the available technology. There are, therefore, two lines of development in their evolution: the extrinsic ones that relate to technical changes, such as improvements in modes of travel and in the means of producing documents through printing, photography, or electronics, and the intrinsic ones that relate to the changing philosophical basis of science, the roles of the scientist in society, and the scientist's modes of organization, education, and interaction. No comprehensive history of scientific communication has been written in these terms. In this chapter I can consider only some of the major developments from the primitive beginnings of science in prehistory to the present, when science has become a highly developed and institutionalized part of our social and industrial systems.

The invention of writing is probably the most important technological advance in communication. Evidence indicates that writing very likely has existed for not more than 5 or 6 thousand years. Before this, society depended entirely on oral transmission from one generation to another to retain its historical and technical heritage. In spite of this limitation, elaborate belief systems developed, as well as an impressive technology involving agriculture, astronomy, and the ability to manufacture both domestic utensils and instruments of war. Oral transmission of ideas persisted even after the invention of writing, particularly in the knowledge of crafts, which was passed on through apprenticeship systems and seldom made a part of the record. The *Académie des Sciences* in Paris recognized this in the eighteenth century when it started a massive effort to collect and publish information on the various crafts of the time, from wig making to coal mining. This effort resulted in an impressive series of monographs in some 103 numbers from 1761 to 1789 under the title *Descriptions des Arts et Métiers.*

The oral transmission of information continued long after the invention of printing, most likely because of a paucity of books. It was the tradition for the professor in the medieval university to read to students from established texts and for them to transcribe the lectures, a system which persists even today. It is difficult to say how much of our knowledge is still transmitted orally and never committed to paper. This is certainly true of some of our social beliefs, and it may also be true of some of the technical aspects of the laboratory. It has been suggested that the scholarship of pre-literate societies, based on oral transmission, tended to be non-analytical, because it is only when ideas are embodied in a static physical form that they can be organized, analyzed, and explored.

Although writing made it possible to fix ideas in a static physical form, some of the problems of oral transmission remained, since manuscripts were disseminated by copying in which errors can easily occur. Now, as in the past, original authors may also fall victim to the editorial zeal of the copyist who may wish to "correct," amend, or even change the text.

One of the earliest scientific manuscripts we possess is the Edwin Smith papyrus, named after the individual who acquired it in Egypt in 1862. It was translated by the famous American Egyptologist James Henry Breasted and was published in facsimile along with a hieratic and English text in 1930. It dates from about 1500 B.C. but is said to be based on older texts going back to 3000 B.C. It includes, for example, explanations of words whose meaning had already become obscure when it was compiled. The Smith papyrus, now in the Library of the New York Academy of Medicine, is one of eight principal Egyptian medical papyri which have been discovered and studied. It is four and one-half inches tall and fifteen feet long, containing a series of surgical cases with a description of the injury and the prescription of a course of treatment. It is obviously only a fragment of the text, because the cases are arranged in the traditional head to foot sequence and stop about in the middle of the body. Each case description is accompanied by a statement about whether the physician will or will not treat it. At first glance this may seem a precaution against charges of malpractice, particularly considering the severe penalties that existed in this period (206). It has been explained, however, as an option which was traditionally exercised by the physician and recognized as late as the eighteenth century.

Another Egyptian medical text of about the same period, called the Ebers papyrus, is about sixty-five feet long and is concerned with nonsurgical problems. One chapter is headed: "Eye diseases treated according to the priest physician as revealed by a Semite of Kepni." This indicates that medicine, even in this early period, was international in scope.

A prominent American surgeon has referred to the Smith papyrus almost with a sense of awe:

> Conceived in ancient Egypt about 5000 years ago, it is the most remarkable book in the entire history of surgery. Compiled by an unknown writer at a time when medicine was magico-religious, when the vocabulary of science had not yet been created and when the first groping steps in inductive reasoning were being taken, this volume is as logical as a modern textbook in surgery. (478)

It is difficult to know the relationships between any of the texts of the manuscript period unless earlier versions exist. This is not true in many cases. It has been estimated that of all the writings produced by the Classical Greeks only about 10 percent have survived even in the form of copies. In addition, many of the copies we have are copies of earlier copies made over a period of hundreds of years to transmit and to replace earlier texts. For instance, the oldest manuscript we have of the work of Hippocrates was written in the ninth century A.D., over a thousand years after his era. The authorship of the whole corpus of Hippocratic writing, which consists of some one hundred treatises, is conjectural. Many of them appear to have been written by his students and their descendants. They are considered to be of the school of Hippocrates rather than by the great physician himself. Others apparently were copied from notes taken by his students at his lectures, a practice which did not stop even with the introduction of printing. Many student notebooks transcribed from lecture notes taken as late as the eighteenth and nineteenth centuries have found their way into library collections, and some even formed the basis for textbooks.

The deficiencies of the manuscript period exist in the process of making copies. The reader has no assurance that a text is the same as that produced by the author or as a copy in the possession of another reader. Texts were transformed in the process of making successive copies and reinterpreted through successive commentaries (117:5). Each copy in a sense constitutes a new edition in which the copyist can exercise and perhaps abuse the editorial prerogatives of correction, expurgation, and emendation, without the author's approval. Manuscript texts thus were subjected to endless alterations, additions, and abridgments. They were frequently brought together in collections of diverse and sometimes disparate items. Transmission of scientific ideas was largely through a process of synthesis, and many encyclopedic works on science and medicine were produced in this way from the Roman through the medieval period. Medieval writers borrowed freely and mostly unquestioningly from other compilers who in turn had borrowed from other sources, until the remote origins are often lost in antiquity. In fact, the ideas of some of the classical writers survive only in the form in which they have been transmitted by these encyclopedic compilers. It may be suggested that these practices have not necessarily disappeared in modern times and that some of our scientific ideas may be transmitted in the same way through textbooks. It may even be valid to suggest that excerpts from published texts which are paraphrased or quoted out of context may be subject to the same abuses as the copied manuscript.

An established trade in books existed even during the Roman Empire. Manuscript copies could be turned out quite cheaply, since they were copied in large shops staffed with slave labor. In some ways this method of production had advantages over our own system of publishing. A "publisher" could turn out a book almost at a moment's notice. The author handed a manuscript to the publisher who turned it over to a staff of slave readers and transcribers. If it was a comparatively short work, it could be ready for distribution in 24 hours without the expense of typesetting or printer's corrections. Estimates have been made that some books in the manuscript period were produced in editions of 500 to 1,000 copies, which is larger than many of the editions produced by the early printing presses. Texts were often borrowed verbatim without any acknowledgment of the source. Other texts maintained their identity, if not their integrity, for a long time. There is, for example, a text of the physician known as John of Bordeaux of which the British Museum alone has twenty-two copies, seventeen in English, four in Latin, and one in Dutch; the earliest is dated 1365, and one was copied as late as the sixteenth century (409).

The use of the handwritten book persisted long after the introduction of printing around 1450, reflecting a cultural lag which often accompanies technical innovations. Objections were raised to printed books on scholarly and aesthetic grounds, although some of them equaled and even surpassed in beauty the manuscript books on which they were modeled. The great advantage claimed for the printed book, that it could replicate identical copies of the same text, was valid only if the original text was properly edited and composed. In their haste to meet market demands, the early printers often did not take care to produce accurate editions. There are even cases cited in this period in which manuscript copies were made from printed books, which may still have been the cheapest way to produce a limited number of copies before the advent of the photocopy machine.

The introduction of printing, however, ultimately caused radical transformations in the nature of scholarship and the transmission of information. Texts could now be made available in identical versions and could be disseminated more widely than had been possible with the manuscript book. It was not only a matter of being able to replicate texts more easily. Authorship could be more readily established and ascertained. A recent writer aptly sums up the impact of printing on scholarship:

> First, it enabled priority of discovery to be established by referring to any copy of a printed text. But this can only be done if the text is accurately dated, and if the author is clearly named. Hence, the accurate establishment of priority of discovery depends on the development of conventions of presentation which record the authorship and date of published works. . . . The possibility of achieving a definitive version of a text, through its publication in a uniform edition of identical copies, also enables research results to be criticized, validated by replication, and incorporated into an accepted body of knowledge by citation. (241:179)

These changes in scholarship were due as much to changes in attitudes toward the writings inherited from the past as to changes in technology. It is said that, up to the Renaissance, no sense of personal property in a piece of knowledge existed. The scholastic method consisted largely of the analysis and citation of authorities. In fact, many genuinely original works were published as those of earlier well-known writers, since the use of an established name conferred an authority which might otherwise be lacking (377:247). Although scholars are finding more and more expressions of original ideas in the Middle Ages, it was largely a period of faith in the ancient authorities. The great German physiologist Helmholtz recognized that this phenomenon still existed in the nineteenth century when he remarked that a large part of our knowledge is still "accepted uncritically and without examination, indeed unconsciously from the past" (214).

It was not only the introduction of printing that helped to usher in an age of improved communication. There were also improvements in modes of transportation and travel, as well as the growth of centralized governments which brought about such changes as better postal services. It is indeed difficult to say whether it was these improvements in communications or the radical changes in attitudes toward the authoritative teachings of the past that brought about the "scientific revolution." Travel in ancient times was one of the primary methods of acquiring new knowledge. There was a certain amount of mobility in the ancient world, but the political unity imposed by Roman rule facilitated greatly the exchange of ideas among scholars. With the breakdown of the Roman Empire, the medieval church took over this function until the growth of the great national states.

Improvements in postal services helped to bring about a great increase in communication among scholars in the form of letters. Out of this tradition of free correspondence among scholars there grew a concept of their belonging to a "Republic of Letters." The word "letters" in that phrase can be interpreted as referring to literature in general or the alphabetic elements of which it is composed, as well as to the letters which they exchanged. The "republic" recognized no political boundaries, and the only qualification for membership was a dedication to the pursuit of knowledge. Vesalius, whose book on anatomy published in 1543 is considered one of the landmarks of modern science, was born in Brussels. No country in Europe, however, can really claim him as its own. He studied medicine in Paris and taught anatomy in the universities of Italy. When it came time to print his great book, he sent the engravings, painstakingly executed in Padua, by caravan across the Alps to Switzerland, a journey of some 3 to 4 weeks, and then followed himself to see the book through the press. He served in the latter part of his life as a physician of the Imperial Court in Spain and died in 1564 while on pilgrimage to Palestine.

Vesalius exemplifies another important change which took place in the Renaissance, the beginning of the coming together of scholars who worked with ideas and words, and craftsmen who worked with things and with their hands. This change can readily be seen by comparing the title page of the *Anathomia* of the medieval physician Mundinus, completed in 1316 but not printed until 1487, with the celebrated title page of Vesalius's *De Humani Corporis Fabrica,* printed in Basel in 1543. In the *Anathomia* the physician is seen sitting on a platform high above an amphitheater where an assistant is performing the dissection, whereas in the *Fabrica* Vesalius is shown standing at the table performing the dissection himself.

Medical books were well represented among the first printed books because, aside from the clergy, physicians represented one of the largest literate classes in Europe. Medicine also was a part of the academic background of the educated man, and even nonmedical readers were interested in medical books as a part of the scholarly traditions. Medical books therefore represented a fairly good investment for the early printers who then as now were motivated to some degree by the need to make a profit. Many of these early books were printings of standard texts which had existed in manuscript form for centuries, but contemporary authors also began to publish their works.

Before the seventeenth century, the literature of scholarship was comparatively static. It was represented by a relatively fixed body of knowledge which, although it was constantly modified by commentaries and recast into new encyclopedic collections, remained of relatively the same order of size as preceding bodies of knowledge. With the emphasis on experiment and personal observation which was introduced by such men as the philosopher Francis Bacon in the sixteenth century, a new element was introduced. The need to report each observation as a separate entity began to take precedence over the tendency to incorporate them into larger syntheses, especially as the means to do so became available through improvements in communications. There was a new sense of urgency which did not characterize the earlier period when scholarship had been regarded as the celebration of old knowledge rather than heralding of the new. This was exemplified by a new pioneering spirit in science in which each scholarly adventurer sought to stake a claim in the new territory being explored. In this environment the ideas of progress and change went hand in hand with the changes in the nature and transmission of information.

The primary medium available to scholars to disseminate new information before the revolutionary emergence of the scientific journal in the latter half of the seventeenth century was personal correspondence. It was personal in that it was usually addressed to an individual, but it was also impersonal in that it served to convey news about observations and experiments rather than personal events. It introduced a new genre which has been called the erudite letter, a letter intended frequently for a broader audience than the one to whom it was addressed. It forms one of the important influences and precedents in the development of the scientific journal, which has today assumed some of these functions in the letters addressed to the editor, as well as in the research and other papers which can be regarded as letters addressed to a relatively undifferentiated and untargeted audience.

Sigerist suggests that the erudite letter continued to serve a purpose even after the scientific journal appeared:

> When a scientist made a discovery in the 18th century, he did not publish it immediately but described it in a letter written in Latin that was sent to some friends abroad. They in turn would discuss these letters with their students and colleagues, would repeat the experiments described and report what their experience had been. After a discovery had been tested in such a way, it might then be published in a monograph, or in the transactions of an academy. (406:12)

The volume of such correspondence in the seventeenth and eighteenth centuries (not all of which took place in Latin, although that was still the preferred language for international communication) is indicated by the size of some of the collections of letters of individual scientists of this period. The Danish physician Thomas Bartholin, who is responsible also for issuing one of the early

medical periodicals from 1673 to 1680, published five volumes of his correspondence during his lifetime and was preparing to publish three more when
his manuscripts were accidentally burned in 1670. The *De Sedibus et Causis
Morborum* of the Italian physician Morgagni published in 1761 (translated into
English as *On the Seats and Causes of Disease* in 1769) was written first in the
form of letters to a friend. Haller's letters fill seven large volumes and include
some 13,000 letters to over 1,600 correspondents. The letters were written in
French, German, English, Italian, or Latin, all languages which he is said to
have used with ease.

To the members of the Republic of Letters international conflicts imposed
no barriers. The Dutch merchant and civil servant who was also the first great
microscopist, Antonie van Leeuwenhoek, continued to send letters to the Royal
Society of London in the seventeenth century, despite the fact that Holland and
England were at war. Much of this correspondence was in fact addressed to the
new scientific and learned societies which were beginning to be formed in the
latter half of the seventeenth century. In some cases both the society and the
correspondence were managed by the same individuals, who became the editors
of the early scientific journals. Henry Oldenburg, a preeminent example of this
class, was the first secretary of the Royal Society of London, as well as the
founder of the *Philosophical Transactions* which appeared in 1665, 3 years after
the Society received its Royal charter. His correspondence, which has recently
been published, filled eleven volumes in 1977 when the editors anticipated that
at least two further volumes would be required (201a). They include letters to
and from the leading scientists of Great Britain and the continent, including
Boyle, Halley, Hooke, Martin Lister, Wren, Huygens, Leeuwenhoek, Leibniz,
Malphigi, Redi, and Spinoza. These letters were to a large extent intended for
the membership of the Royal Society with whom he shared them. They provide
a fascinating glimpse into the scientific life of the seventeenth century. The letters
were recorded in the Record Books of the Society, but many of them also found
their way into Oldenburg's new journal where he could share them with a wider
readership.

Among the innovations which preceded the introduction of the scientific
journal were the publications which established predictable channels of communications and created what we can call "periodicity in print." They came in
various forms such as the annual book catalogs which were published for the
various European book fairs shortly after the intoduction of printing. There
were also manuscript and printed newsletters which were issued for a select
clientele on a regular basis. Other important forms were the calendars and
almanacs which provided an annual vehicle not only for lists of phases of the
moon and saints' days, but also such things as agricultural and health information
and advice. Printed newsletters could also serve as vehicles of scientific information. It is alleged that in 1608 Galileo first learned of a device for seeing
objects at a distance from a twelve-page Dutch newsletter. Appended to an
account of the visit to Holland of an embassy from Siam was a short note about

the use of a combination of lenses which made distant objects appear closer (118).

Newspapers were widespread by the middle of the seventeenth century. The first French newspaper was issued by the crusading physician Theophraste Renaudot (1586–1653), who issued the *Gazette* under the auspices of the government in 1631. His sensitivity to the condition of the poor led to other innovations, including a public health clinic which provoked the bitter enmity of the medical establishment represented by Guy Patin and the Faculty of Medicine of the University of Paris. The early newspapers carried news of interest to the scientific community, but it was not until 1665 that the first two journals which can be regarded as true organs for the dissemination of scientific ideas appeared. The *Journal des Sçavans,* as it was then called, appeared in Paris in 1665. The *Académie des Sciences* formed shortly after in the same city had no formal association with the *Journal* although news of its activities did appear in its pages. Although the *Journal* reported scientific discoveries such as Jean Denis' curious experiments with the transfusion of blood and other scientific news gleaned from its correspondents, it can be considered more a general literary journal than a scientific one. It was primarily concerned with reviewing books in every field of knowledge, except for those on religion and politics, subjects which the editors treated very gingerly. The *Journal* was soon reprinted in Amsterdam and other places and became a model for literary journals throughout Europe.

The *Philosophical Transactions* which also appeared for the first time in 1665, on the other hand, devoted itself almost exclusively to the concerns of the Royal Society of London with which it was closely associated through its first editor, Henry Oldenburg. The *Transactions* did not, however, acquire formal sponsorship by the Society until almost a century later. The *Académie des Sciences* pursued another course by publishing its proceedings in quite a different format which in turn became a prototype for similar publications by other learned societies on the continent. Both the *Philosophical Transactions* and the *Journal des Sçavans* celebrated their three hundredth anniversary in 1965 although in quite altered form. The *Journal des Savants,* as it is known today, is a general literary journal, and the Royal Society uses the *Transactions* now as a medium for the publication of two series of highly specialized monographs, one in mathematics and physics, and the other in the biological sciences.

Modern commentators assign a high importance to the appearance of the journal in the history of science. Ziman puts it most emphatically: "The invention of a mechanism for the systematic publication of fragments of scientific work may well have been the key event in the history of modern science" (475). Cole and Eales say pretty much the same: "It may, in fact, be claimed that science could not have made the advance that it has but for the recognition of the periodical as the most convenient and efficient method of encouraging research" (79:588).

In many ways the scientific and learned journals of the seventeenth and eighteenth centuries did not resemble very closely the research journal we know today. Borrowing from one journal to another was a common practice. For the most part the source was acknowledged, but frequently no indication of origin was given. Sometimes the borrowed information was paraphrased and commentary was added. In this respect they were very much like the newspaper which provided a model. They did not worry about redundancy because they regarded themselves principally as news media. Duplication of information becomes a problem only when the recipient has access to both sources. This poses an interesting difference between the newspaper and the scientific journal in that the quality of a newspaper is improved in the degree that it includes information from other media, since it usually provides a unique source for a particular reader. A scientific journal, on the other hand, particularly one publishing original research, is proscribed from this practice. In the seventeenth and eighteenth centuries, journals were not as closely differentiated from books as they are today and were frequently reviewed in the same manner. This reflected the fact that in some instances a large part of the content of a journal issue could be attributed to the editor, who sometimes referred to himself as the author. They were also characterized by short duration; many did not last more than 2 or 3 years. This is a corollary of the fact that many of them appeared under the auspices of a single individual.

Specialized journals began to appear very early in the history of the scientific journal, although much science continued to be reported in the general scholarly journals. Among the first specialized journals was one edited or written by another controversial French physician, Nicolas Blegny (1652–1722), who is regarded by some historians as a rather unsavory character although he is also credited with outstanding work on hernias in the seventeenth century and other interesting social and medical innovations. His *Nouvelles Déscouvertes sur Toutes les Parties de la Médecine* appeared in 1679 in the form of letters to a provincial physician. It can also be considered an example of how scientific journalism can be used for self-aggrandizement as well as for the dissemination of information. Blegny took care in some of his discussions of new drugs to indicate where they could be obtained and was not constrained from speaking well of a new kind of hospital he had opened (262).

The earliest specialized journals were medical because physicians represented one of the largest organized groups interested in the new science. They were also used to foster the intellectual and social status of a specialty. The *Académie de Chirurgie* organized in Paris in 1731 in recognition of the surgeon's claim to academic standing began to publish the *Mémoires pour les Chirurgiens* in 1736. Specialized journals in other disciplines did not begin to appear until later in the century. The German chemist Lorenz von Crell (1744–1816) published his *Chemisches Journal* from 1778 to 1781 and followed it with other journals in chemistry. In England William Curtis (1746–1799) began his *Botanical Magazine,* which is still avidly sought by collectors today for its beautiful colored plates, in 1787. Before the end of the century, there was even a journal in psychiatry, or what the Germans called "knowledge of the soul," the *Magazin der Erfahrungsseelenkunde,* edited by Carl Philipp Moritz from 1783 to 1793.

The predominance of independent journals, that is, those not sponsored by scientific or learned societies, continued through the middle of the nineteenth century, but society sponsorship became more prevalent as time went on, and today the situation is quite the reverse, with most journals sponsored in some way by societies. The famous Danish physician Thomas Bartholin (1616–1680) published his *Acta Medica et Hafniensa* from 1671 to 1680 while he was on the faculty of the University of Copenhagen, but it had no official connection with the school. The early scientific societies nevertheless had a strong influence on the development of scientific journalism, since they both developed as a response to the same intellectual and social needs.

Societies also developed another form of publication similar to but different from the journal. The Royal Society of London contented itself with the *Philosophical Transactions* as a medium for disseminating news of its activities. The *Académie des Sciences* in Paris, however, developed its own unique medium by publishing its proceedings in the form of a series called *Histoires et Mémoires.* These in a sense form an archival record of the activities of the society as well as a repository of the most important papers which had been delivered in person or communicated to the society. They were often, however, published sporadically or long after the papers were presented. The *Histoire et Mémoires* of the French Academy formed a model for the scientific and learned societies which proliferated in the French provincial cities, throughout Germany, and elsewhere in Europe. To a large extent society proceedings provide a more accurate prototype of our modern scientific journals than the independent journals which were published at the same time. Some of our current editorial practices such as peer review had their origins in the methods these early societies devised for accepting communications for publication.

It is alleged that the Royal Society of London was the first to introduce "the concept of refereeing" in the middle of the eighteenth century by setting up a committee to review all papers before they were published in the *Philosophical Transactions* (48). There were, however, a large number of antecedents to this practice. Oldenburg had the responsibility of screening communications for presentation to the Society, but after the papers were read, they were "ordered to be reviewed by several of the Fellows" (202:70). The *Académie des Sciences* in Paris early in its history established select committees to determine whether or not a member could publish under its auspices. The peer review process almost as we know it today is described in the preface to the French edition of the *Medical Essays and Observations* published by a "Society in Edinburgh" in 1731. The papers submitted, it informs us, are distributed according to their subject content to those members of the society who are more versed in these matters for their review. It also specifies that the identity of the reviewer is not made known to the author, an early example of the controversial anonymous reviewer (131). The *Société Royale de Médecine* soon after its institution in 1776 inaugurated a system by which two of its members were delegated to examine each paper submitted to the society and to provide the other members with a summary and critique (205:121). Validation of scientific work through a process of review and discussion was in fact a major function of the early scientific societies.

Reasons for the formation of scientific societies may be sought in the generally gregarious nature of humanity, but equally important was the failure of the contemporary university to respond to the ideas stimulated by the scientific revolution. These organizations of individuals interested in the new science were sometimes called "invisible colleges" because they carried on some of the traditional functions of the universities outside their confines. This term has been adopted in modern times to describe groups of scientists working on a common problem who are known to each other and who establish their own communication networks. Today's meaning differs from that of the seventeenth-century "invisible college" which was to a large extent an open society to which individuals were admitted on the basis of their interests and aptitudes, rather than the inner circle which is suggested by the modern "invisible college."

Some of the early scientific societies made contributions of papers a condition of membership. Many of them also gave prizes for essays submitted on questions proposed by the society which were widely publicized through the general and special periodicals. These prize essays in many ways can be considered closer precursors of the modern scientific paper than any other eighteenth-century publication format. They can to some extent be regarded as a form of sponsored research. This is clearly evident in the prizes awarded by such organizations as the Royal Society of Arts, originally the Society for the Encouragement of Arts, Manufactures and Commerce, which from its origins in 1754 offered prizes for essays on such subjects as the improvement of the production of wine and the development of a more efficient ship's pump. The Society of Arts and Science in Utrecht proposed a question in 1783 which has a decidedly modern inflection: What are the causes for the increase in nervous illnesses, and which causes lie in nature and which in the mode of life. They offered a prize of thirty ducats and expressed their willingness to receive papers in Dutch, French, or Latin (11:49). The venerable *Faculté de Médecine* in Paris, stimulated at last by the activities of the newly organized *Société de Médecine* to publish an account of its public meetings, in 1778 announced that awards had been made for essays submitted on the subject of the treatment of miliary fever in pregnant women. Five essays were regarded as of special merit, but only two were elected to share in the prize (397:13–17).

The role of the prize essay in the history of science has not yet been fully explored. Prizes are still awarded for essays or scientific papers today, but they are usually on subjects of the author's own choice, whereas the eighteenth-century prize was on a subject announced in advance and open to international competition. Contestants submitted their papers anonymously to the society's secretary with some kind of device to identify them later, usually a Latin motto. They resemble to some extent the dissertation, an academic tradition which was centuries old. In fact, the *Akademie der Wissenschaften* in Berlin published a collection of their prize essays in 1748 under the title *Dissertations qui a remporté le prix*. The societies were careful to preserve the anonymity of the author in order not to be influenced by reputation. The mathematician Daniel Bernoulli nevertheless was awarded the prize by the *Académie des Sciences* ten times in the eighteenth century. The attempt to preserve anonymity was in keeping with their efforts to eliminate personal bias not only in the judgment of research results but also in the observation of nature. It is interesting to note that this period, which has been characterized as one of the depersonalization of science, was accompanied by a greater emphasis on priority which gave rise to a number of acrimonious disputes. It reminds us of Merton's discussion of the clash between the social norms of science and the personal needs of the scientist, and Kuhn's comments about the "essential tension" in science.

Almost half of the learned and scientific journals which appeared in the eighteenth century were published in German. This phenomenon has been attributed to a German quality which has been described as *schreiblustigkeit,* or joy in writing, and to the highly decentralized nature of the German-speaking states at that time, in which each provincial state created its own journals. It also reflects the parochial nature of many of these publications because German was a language not widely used outside of its sphere of dominance. Latin had been the dominant scholarly language until about 1600, when vernacular languages like French and English began to be used more frequently in scholarly writing. In the eighteenth century French assumed ascendancy, but this was as much a result of social and political factors as scientific ones. French retained its dominance in the first part of the nineteenth century because of the outstanding contributions of its scientists and physicians, but gradually gave way to German until at the end of the century it was almost mandatory for any serious scientific worker to read German.

This phenomenon has been said to have been produced by the great increase in productivity among German scientists. One of the strong reasons for this phenomenon, says the sociologist Ben-David, was the decentralization of the German universities which, like the proliferation of German journals, was a consequence of the political decentralization which existed when they were created. The existence of a large number of academic institutions made possible a high degree of mobility among students and teachers. This in turn produced a condition of scientific competition which was not duplicated elsewhere. It was accompanied, Ben-David says, by a change from the dominance of *Naturphilosophie,* which was a philosophic way of looking at natural phenomena, to a greater emphasis on experimental science. This in turn led to the recognition of the university as a seat of scientific research and to the creation of scientific institutes, all of which figured in the competition for faculty. American universities with graduate programs in the sciences tended to model themselves on the German universities in this period. This change, Ben-David concludes, created some of the conditions under which American science assumed ascendancy in the twentieth century, until now English rather than German is the predominant language of the scientific world (30).

A large number of the journals which had their origins in the latter half of the nineteenth century bear the names of outstanding scientists of this period like *Wilhelm Roux' Archiv für Entwicklungsmechanick der Organismen,* which began in 1865. These were not merely honorific titles because in most cases the journals were inaugurated by scientists who took personal editorial responsibility for them. This tradition of concern for the literature by eminent scientists has an old tradition and is, indeed, a hallmark of professional status. The Swiss physician, physiologist, botanist, and poet Albrecht von Haller (1708–1777) has been called by William Osler "the greatest bibliographer in our ranks" (345). Early in his life Haller developed the practice of systematically reading and abstracting the literature. This activity culminated in his eight-volume textbook on physiology, *First Lines of Physiology (Primae Linae Physiologiae)* published in Göttingen in 1747, and in his monumental bibliographies in anatomy, surgery, medicine, and botany. One biographer credits him with having written 12,000 book reviews for the *Göttinger Gelehrte Anzeiger* during his tenure at the university from 1745 to 1777. This would average out to about 400 reviews a year, a formidable achievement even for a genius like Haller, considering the other things he was doing like developing a theory of tissue irritability. Fortunately for our sense of credulity, a later scholar found that his biographer had been overcome by admiration and added an extra zero to the 1,200 reviews with which he had originally been credited (383).

Many leaders in biology and medicine began journals in this period to create outlets for their own work and for the work of their colleagues and students. The German anatomist and physiologist Rudolf Albert von Kölliker (1817–1905) founded the *Zeitschrift für wissenschaftliche Zoologie* in 1848 and served as its editor for half a century. Rudolf Carl Virchow (1821–1902), who was probably the best known German physician of his time and who, as the founder of cellular pathology, contributed greatly toward the scientific basis of modern medicine, began the *Archiv für pathologische Anatomie und Physiologie* in 1845 and remained its editor until his death. It became known as *Virchow's Archiv,* but this did not happen until after his death. This is also true for other outstanding journals of the period like that of Albrecht von Graefe (1828–1870), who made numerous contributions in ophthalmology and edited the *Archiv für klinische und experimentelle Ophthalmologie;* Felix Hoppe-Seyler (1825–1895), who occupied the first chair of physiological chemistry in Germany and edited the *Zeitschrift für physiologische Chemie;* and Carl Gegenbaur (1826–1903), the anatomist who was responsible for the *Morphologischen Jahrbuch.* It is interesting to note that these journals today publish largely in English and that some of them have even acquired English titles.

It has been estimated that the number of all periodicals increased from about 900 in 1800 to almost 60,000 in 1901 (61:102). During this period the scientific journal changed from what was primarily a news and book review medium to a vehicle and repository for scientific research. It did not lose any of its old functions in the process of adding new ones, but it did go through a process of specialization and diversification which has produced the complex and cumbersome system we know today.

One of the consequences of this growth has been the increase in secondary media, that is, publications which are based on antecedent publications and present guides or digests (another biological term) to enable scientists better to gain access to or to control the burgeoning literature. Efforts to provide summaries or syntheses of the current state of knowledge (to present state-of-the-art reviews as we say today) go back to the beginnings of recorded knowledge. The great encyclopedic compendia of the classical and medieval periods, like the Roman Pliny's (23–79) *Natural History* and the Franciscan known as Bartholomew the Englishman's (c1260) *On the Properties of Things,* represent efforts of this kind. With the proliferation of journals in the eighteenth century, anxieties began to be expressed about the difficulty of gaining access to the literature. One of the new techniques introduced at this time was the abstract journal. A journal published in Mannheim in 1760 describes the function of this new medium precisely in its title: *Journal des Journaux; ou Précis des Principaux Ouvrages Périodiques de l'Europe* (Journal from Journals; or a Summary of the Periodical Works of Europe). The new medium apparently met a real need because the number of active abstract journals in all fields increased from nine in 1790 to 249 in 1920 (294:136). Germany dominated in providing these services in the sciences up until World War I when many of these publications were disrupted and not resumed. The *Chemischer Zentralblätt,* which began in 1830, established a model for a series of *Zentralblätter* in various disciplines, some of which lasted for only a few years but many of which continued for decades.

Other efforts to organize and synthesize the literature took the form of progress reports and annual reviews which took such names as *Jahresbericht* (annual report) as in the *Jahresbericht über die Leistungen und Forschritte im Gebiete der Ophthalmologie,* which began in 1870. They also began with terms like *Jahrbuch* (yearbook), *Ergebnisse* (results), and *Fortschritte* (advances), all of which have their cognates in English and other languages.

The first efforts to index the journal literature took place almost immediately after the first journals appeared. An obscure Flemish bookseller named Cornelius a Beughem published an index to the *Journal des Sçavans* in 1683 under the title *La France Scavante.* Like all Gaul it was divided into three parts, a chronological part in which the contents of each issue were listed by date of publication, an author index, and finally a classified subject index. It is interesting to note that the *Current List of Medical Literature,* which was published by the National Library of Medicine from 1941 to 1959, was in almost the same format. Beughem followed his single-journal index with an index to a larger number of journals which included the *Philosophical Transactions* in its Latin edition, the *Acta Eruditorum,* a Latin equivalent of the *Journal des Sçavans* published in Germany, and seven other journals. Many other indexes either on a current or retrospective basis appeared in the eighteenth and nineteenth centuries. They had their culmination in a large retrospective index which the Royal Society began to publish in 1858 to cover the scientific literature in all disciplines published since 1800, under the title *Catalogue of Scientific Papers.* Later in the century, the National Library of Medicine (then the Library of the Surgeon General's office) began to publish the monumental *Index Catalogue of the Library,* which although primarily devoted to medicine is of interest in all the life sciences. It appeared in five series from 1896 until it was discontinued in 1961, when it was decided that it could not keep pace with the growth of the medical literature. These publications are still of great use for the historian of science or the investigator looking for early antecedents of scientific ideas. They represent precursors and prototypes of a host of publications in all disciplines in the life sciences, many of which are now being mediated by a powerful new tool which has only recently become available to provide a new technology for dealing with some of the problems of control of and access to the literature, the computer. The social and intellectual factors which produced these publications, however, cannot be ignored in considering the direction in which the evolution of the media of scientific communication is taking us today.

Chapter 3

Varieties of Information Sources: Primary

Over the course of centuries, we have evolved a large number of information sources of various kinds. An indication of the extent to which the industries devoted to information services have grown is a directory of current information systems and agencies published early in 1981 (129). Over 2,500 agencies were identified at that time. By July 1983, over 300 new ones had been added to the file. In addition to these agencies, a large number of other sources of information are published each year, including textbooks, manuals, directories, and guides to the literature. These sources do not lend themselves easily to classification or analysis, because few fit neatly into any category. They can, however, be broken down into broad groups, which may assist an inquirer in making an appropriate choice of an information source.

The simplest and at the same time the most difficult method would be to classify them into good sources and bad sources. This involves a judgment not only in terms of the quality of the source but its appropriateness to meet a particular information need. Other qualitative judgments would, of course, be based on an assessment of the reputation and authority of the source and past experience in using it. The following three aphorisms by an experienced literature specialist state the case well:

1. The literature is a tool just as surely as is a spectrograph or a microscope.
2. A tool is as good as its user.
3. A professional is constantly improving his tools and his technique for their use. (399:2)

An old laboratory joke states that, in introducing a new machine, when all else fails you should read the manual. Contrary to this apocryphal view, the more you know about an information tool the more effectively it can be used.

Among the things which you should know about any information source are those which can be considered under the following headings.

1. *Scope.* This factor concerns not only the subject areas covered, but also the particular orientation chosen by the producer, for example: Is it directed primarily toward practitioners or research investigators? Or, if it is a directory of research awards, does it cover private as well as public agencies?
2. *Coverage.* Consideration of this factor will determine whether the source is used for current (recent) or retrospective (older) literature searches. Included here can also be the languages and countries covered. Most bibliographies, whether current or retrospective, have a language bias depending on the country in which they were compiled or the policies of the producers. Coverage also involves such things as the time span of the publication, such as the beginning and closing dates for a review, which cannot always be determined from the publication date.
3. *Currency.* This concerns how up to date the publication is and is also related to its time span. It is also concerned, particularly with current indexes, with the lag between the time an agency receives a document and the time it appears in its printed publication or computer-based online service. The *Index Medicus,* a publication we will consider later, claims an average lag of 6 months from the time a journal is received for indexing until it appears in print in the index. This does not take into consideration the late receipt of a journal or the fact that it may be in "exotic" languages, usually defined as those languages not accessible to a member of the current indexing staff.
4. *Format.* Formats vary in the way information is arranged, for instance, whether in a classified or alphabetical order, and in the nature of the indexes and other kinds of access they provide. Physical form can be considered here also, especially with the increasing number of information sources that are appearing in "non-print" formats such as microforms and electronic versions.
5. *Authority.* The question of how authoritative the information source is, that is, how much can it be relied upon, is probably one of the most difficult to answer. The answer is based to a large extent on the reputation of the sponsoring agencies, the authors, editors, and compilers, and the reputation it has acquired in use. Science is built upon skepticism, but to some extent it is also built on trust.
6. *Subject Access.* This factor relates more to access tools like indexes, abstracts, and data collections, but it can also enhance or diminish the value of a textbook or monograph. I will discuss subject indexing methods and vocabularies in Chapter 8, because they have an important influence on the effectiveness of any particular source.
7. *Author, Name, and Other Access.* Here we are concerned with such things as the inclusion of secondary authors, names of sponsoring agencies, and other access keys in the indexes.

These are all factors which can be considered in determining whether an information source is good or bad, that is, appropriate or inappropriate for meeting a particular information need.

Information sources in science are more conventionally classified into what are called primary and secondary literature. The term primary is used for literature which reports original research or other new ideas or opinions for the first time. Such reports can appear in journals, monographs (books), report literature (usually individual documents reporting funded research), patents, or dissertations. Secondary literature is that which is derived from the primary literature in order to organize it, to provide access to it through abstracts and indexes, or to summarize it in the form of reviews. It might be better, therefore, to use the terms original and derivative to refer to these two classes of literature, particularly since primary is sometimes used to describe the level of complexity of the materials. Sharp divisions are difficult to make because publication media such as journals are frequently mixes of both original and derivative materials. Also, it is often not easy to distinguish between what is original and what is derivative. A review is derivative in the sense that it is a response to the published literature, but it may also be original in that it provides a new point of view, a new evaluation, or a clarification. Textbooks may be considered primary because they present a synthesis of a particular subject which is highly original. They are also primary in the sense that they present a starting point in approaching a new subject. They are secondary, however, in that they represent state-of-the-art reviews derived from the literature.

The question of primary (original) versus secondary (derivative) sources of information is one that cannot be easily resolved for monographs and journals, as we have seen. It may even be true that originality is something that exists only in the eye of the beholder. Information that is new to the recipient is original because it has not been seen before, even as the morning newspaper reports original news to the reader even though it has been repeated in newspapers across the country. "Original news" is an example of a tautology; what is new to the reader is also original. The reader's concern must be in the credibility and reliability of the source, and the auspices under which the information appears. Gaston's comments seem highly appropriate here:

> ... Originality has both a substantive dimension and a time dimension. A contribution that replicates or explicates earlier work may be original, but it probably is not so important for science as the first work reported. Thus originality is a variable that is not dichotomous, original or unoriginal, but is bound by some continuum from truly original to original only in some small way. . . . The independent discoverer may have used as much creativity and hard work as the first scientist, but even, if the second announcement comes only shortly after the first, it is still announcing something already known. (177:11)

The term secondary in relation to information sources has a narrower and more selective meaning. It refers to those publications which do not stand by themselves but are based on other publications for which they serve as location devices or surrogates.

In this and the following chapter I deal with the types of information sources which fall into these two major classes. Although I cite a few examples of each type, it must be clear that they represent only a small fraction of the many information sources that are available in each discipline. Some of them have a long history, but new ones are added every year, and old ones are frequently updated and revised. I will be able to consider some of the more important ones, but you can turn to other sources to find those that are most appropriate for your needs. The catalogs of research libraries, information specialists, and reference librarians in research libraries are good sources. There are also numerous guides to the literature of various disciplines. I have listed some of these in the Appendix. It is also important to emphasize that much information is acquired from other than formal sources, from the casual reading of journals and other publications that come from manufacturers of laboratory instruments, media, and pharmaceuticals, from casual or organized encounters with colleagues, such as those that take place at conferences and congresses.

Conferences and Congresses

One indication that oral discourse is still a flourishing method of transmitting scientific information is the proliferation of scientific conferences and meetings in the past few decades. They range in size and purpose from such lofty and exclusive invitational meetings of certified experts as Arthur Koestler describes in his novel *The Call Girls* (254) to the huge gatherings such as those of the Federation of American Societies for Experimental Biology which attract thousands of scientists from a broad spectrum of disciplines. It has been estimated that at least 10,000 conferences are held every year and that about two-thirds of them publish proceedings (165). The growth of international conferences alone is shown by a study which calculated that from three in 1853 they grew to over 100 in 1909 and to at least 2,000 by 1953 (245). These meetings are sometimes also called congresses, institutes, symposia (originally a convivial gathering for drinking and conversation), workshops, or clinics. They all conform to the old definition of conference as an assembly of individuals for the purpose of presentation and discussion of ideas, and "taking counsel" as the *Oxford English Dictionary* puts it. They are organized, says another source, "to present current status and progress reports on a specific subject" (388). They also provide a forum in which the results of research in progress can be presented and discussed.

One tongue-in-cheek comment was that conferences provide "a place where conversation is substituted for the dreariness of labor and the loneliness of thought" (147). The information presented at meetings is sometimes regarded as "soft" in the sense that it may not have been carefully reviewed before acceptance and in that it represents preliminary communications which have not been admitted into the archival record. Papers presented at meetings do not always reflect what has been presented in advance abstracts, because of the stringency of submission deadlines which sometimes require that abstracts be submitted while the work is still in progress. It is more and more frequently forgotten that the submission of an abstract to the organizers of a meeting represents a commitment to present a paper whether or not the work has produced useful or significant results. Conference papers are unfortunately sometimes also a kind of "pièce d'occasion" put together for the requirements of the meeting rather than being prompted by the need to present new ideas or research results. Paramount among the useful purposes served by conferences is the opportunity for face to face encounters with one's colleagues.

Practitioners and investigators can generally learn about the meetings they may wish to attend from announcements in the specialty journals and newsletters of the societies to which they belong. The journal *Science* updates a list of meetings quarterly in its news pages. It provides the meeting dates for the American Association for the Advancement of Science, for which the journal is an organ, and also for its many affiliated societies and other key scientific organizations. The announcements are generally printed about a year in advance, but only meeting dates are supplied. The *Journal of the American Medical Association* includes among the "reference directories" which appear in its weekly issues in three separate reports for each volume, that is, four times a year, a list of state medical associations and their meeting dates, along with dates for examinations and licensure in various medical specialties, a list of scientific meetings in the United States, and finally a list of meetings outside the United States.

There are also several separate publications devoted to listing dates of scientific meetings, such as *Scientific Meetings,* which appears quarterly and has been published by Scientific Meeting Publications since 1957. It provides an alphabetical list of associations sponsoring meetings along with their addresses and a chronological list covering meetings about a year ahead with places and dates. More complete information is supplied by a series called *World Meetings* published by the Macmillan Company. It appears in four series: *United States and Canada, Outside United States and Canada, Social and Behavioral Sciences, Education and Management,* and *Medicine.* These lists supply information on sponsor, subjects covered, estimated attendance, whether contributed papers are solicited or only invited papers are considered, and deadlines for submission. They include indexes by keywords taken from the meeting title and name of organizational sponsor, and indexes by meeting date, deadlines, and location. Meetings are listed sometimes as long as 2 years in advance depending on whether the publisher has been notified. With such a list in hand one is in a position to satisfy almost any kind of zeal for scholarship, as well as almost any urge to travel.

Some organizations such as the Federation of American Societies for Experimental Biology publish abstracts of papers to be presented well in advance of their meetings. One of the shortcomings, as I have pointed out, is that some of the papers presented at the meetings may not bear much resemblance to the abstract. Another problem, particularly in periods of financial stringencies when travel budgets are restricted, is that some of the announced papers may not be

presented. Some papers presented at conferences or congresses are never published at all. In addition, other prestigious conferences do not publish abstracts
or papers presented in order to promote a freer exchange of ideas (i.e., some
Cold Spring Harbor conferences, Gordon conferences). In a study of 383 contributions made to four U.S. conferences in the 1950s, it was discovered that
almost half never appeared in print, and the other papers appeared anywhere
from 1 to 5 years later (275). The published proceedings of many conferences,
nevertheless, provide a useful way to acquire state-of-the-art reviews and notices
of research in progress. They sometimes also supply an account of the discussion
generated by the reports, and bring together a body of information about a
specific topic or research area that is not readily available elsewhere.

Conference papers can appear in any one of a number of formats:

1. As "preprints," copies prepared for distribution prior to or at the conference
2. In single- or multiple-volume monographs numbering anywhere from one to
   ten or more, which include all the papers presented at the conference, as well
   as the ceremonial addresses and reports of other amenities
3. As single volumes with an editor and specific title
4. As supplements or special issues of an existing journal, frequently that of the
   society responsible for the meeting
5. As individual articles published in separate and diverse journals.

It is hard to know how much of the journal literature is a result of papers
presented at conferences because they are not always so identified. Conference
papers which appear in journals that are covered in the standard indexes and
abstracting media are probably easiest to find. The proceedings of certain recurring conferences such as the *Cold Spring Harbor Symposia on Quantitative
Biology* are published as part of a series under a common title, and the papers
are cited like those which appear in ordinary journals. The New York Academy
of Sciences publishes all its many symposia in its *Annals,* which have been
published since before the turn of the century. In 1980–1981, the proceedings
of almost fifty conferences held under its auspices were published, covering a
spectrum of basic and applied sciences and subjects ranging from "Micronutrient
Interactions" (vol. 355, 1980) to "Victorian Science and Victorian Values: Literary Perspectives" (vol. 360, 1981). Conferences and symposia which appear
routinely in certain series such as *Bibliotheca Anatomica* are also covered by the
standard abstracting and indexing media.

The sharp distinction made between journals and monographs has made it
difficult to locate conference papers published in other than journal formats.
This has been remedied to some extent by the efforts some of the abstracting
and indexing media make to include what they call "multi-authored works,"
that is, books which contain contributions by many individual authors. The
Current Contents series lists the contents of selected conferences in the front of
each issue. *Biological Abstracts/RRM* (formerly *Bioresearch Index*) lists published conferences in the book synopses which preface each issue, and since 1977

Science Citation Index has been analyzing separately published proceedings in the same way as selected multi-authored books or collections of papers. From 1976 to early 1981, *Index Medicus* included citations to conference papers published in nonserial monographs, but discontinued it because it interfered with the ability to incorporate all the new high-quality journals which were making demands on the system (19).

In addition to the sources cited above there are publications devoted exclusively to indexing conference papers:

- *Directory of Published Proceedings* (Harrison, N.Y.: InterDok Corp.) appears in two series, one devoted to *Science, Engineering, Medicine and Technology*, issued ten times a year, and the other to *Social Sciences and Humanities*, published quarterly. The conferences are listed chronologically by conference dates, along with any publication which may have resulted. It is interesting to note a 2- to 3-year difference in some cases between the conference and publication dates. Indexes by names of editors, and by keywords derived from the names of the sponsoring agencies and conference titles, are included.
- *Conference Papers Index* is a monthly publication by Cambridge Scientific Abstracts in Washington, D.C., which has been published since 1973 and is said to have almost 1 million items in its database, which is also available online. It lists papers presented at meetings whether or not they are published, although publications are listed when available. The printed version is arranged in seventeen sections under broad topics, many of which are of interest to the life scientist.
- *Proceedings in Print* was originally published by the Special Libraries Association but is now published bimonthly by Proceedings in Print Inc. with an annual cumulative index. Only papers in print are included, as well as those published in journals.
- *Index to Scientific and Technical Proceedings* has been published since 1978 by the Institute for Scientific Information and is the most comprehensive and in-depth index available to this type of publication. In 1982 the Institute planned to cover 3,000 of the most significant published proceedings representing some 100,000 papers. It was estimated that 30 percent of the papers fall within the life sciences, which includes psychology with a physiological orientation. If we include the 8 percent each for clinical medicine and for agricultural, biological, and environmental sciences, the total comes up to almost half. It is published in seven sections:
  1. Category index, under major headings such as "Immunology" (eleven proceedings publications were listed under this heading in the August 1982 issue).
  2. Contents of the proceedings publications listing all the papers by author and title along with procurement information.
  3. Author and editor index.
  4. Sponsor index.
  5. Meeting location index, by country and city where held.

6. Permuterm subject index based on the title words in the individual papers.
7. Corporate index in two parts, by location and by name. This index is also available online and can be searched in similar ways to other ISI databases.

Journals

Journals, or periodicals as they are also called, are publications with a uniform title which are issued at more or less predictable intervals anywhere from once a day to once a year. They are primary in that they constitute the principal method of transmitting original, timely, and new information to the community of life sciences practitioners and investigators. They are also primary in that most reports of original research are disseminated in this way. In a study of the literary output of one productive research laboratory, it was discovered that 50 percent of the written communications were published in journals, 20 percent as abstracts for conferences, 16 percent as solicited contributions which appeared in a variety of formats, and 14 percent as chapters in books. The analysis also revealed that 55 percent appeared in specialized journals related to the laboratory's research emphasis, 5 percent were written for a lay audience, that is, for such publications as *Scientific American,* and 27 percent were addressed to scientists who were working outside the laboratory's primary area. Review articles accounted for another 13 percent. Of the specialized journal articles, three journals accounted for 38 percent (272:12). It is interesting to note that, except for those published in books, the remainder appeared in journals and periodicals of one kind or another. Another measure of the importance of the journal as a communication medium is the fact that 80 percent of all the references in the scientific literature are to journal articles (172).

Estimates of the total number of scientific and technical journals existing in the world today range anywhere from 30,000 to 100,000, which leads one to suspect that there may be differences in the way the data are gathered. Moreover, most observers agree that only a small percentage of these journals contain information significant to the advancement of scientific research and practice. It has been estimated that not more than 10,000 are significant enough even to consider for indexing. Even this number is too large to be handled comfortably and economically. Accordingly, a solution has been to select only those which tend to yield the highest number of useful articles. This is also helped by the fact, as we shall see, that the most cited articles tend to cluster in a small number of journals.

Overall, there has been no diminution in the appearance of new journals every year. The situation resembles a huge intellectual smorgasbord, from which it is more and more difficult to select an appropriate diet. To some extent it may be said that the more journals that are brought into existence the less communication takes place, because of the higher degree of scatter of information that inevitably occurs. The journal *Nature* for two consecutive years published a review of the outstanding new scientific journals which had appeared in the previous years. In 1981 of the 124 journals reviewed more than half could be regarded as concerned with the life sciences (263). An accompanying editorial pointed up the dilemma: "There are always more journals to be read than can be read, —but there are never enough outlets for contributions to the continuing torrent of discovery" (291). The review the following year covered only sixty new journals, and all but sixteen could be considered as directed toward the life sciences (68).

Journals have exhibited a growth not only in numbers but also in the average number of pages they publish each year. In a study of thirty-seven primary journals in biochemistry from 1968 to 1977, it was discovered that the total number of papers increased by 59 percent, from 9,060 to 14,418. Some of this increase was accounted for by the introduction of new journals, an average of one per year for the 10-year period. However, the average number of pages per year published by these journals increased by 19 percent in the same period, from 335.6 in 1968 to 389.7 in 1977 (174).

Scientific journals encompass such a wide variety of types that any effort to classify them must result in a limited and distorted picture. "Scientific journals," says Maeve O'Connor, "whether sponsored by learned societies or by commercial publishers, vary from the ultraspecialized to the general, from the rich giant to the impoverished dwarf, and from the academically detached to the politically engagé" (332:14). Although we do not generally think of scientific journals as being "politically" oriented, biases can sometimes be observed. Distinguished journals such as *Lancet,* which was inaugurated in 1823 by Thomas Wakeley, began as efforts at medical reform. The *Chinese Medical Journal* during the period of Mao's leadership began even its scientific articles with references to the views of the Chairman. Occasionally, even political and economic ideologies can find their way into the scientific literature as suggested by this translation from a Russian journal published in 1952:

> . . . In the study of foreign journals (capitalist) it is necessary to take into consideration that they not infrequently publish material of little scientific or practical interest: repeating long known and out of date theories, methods and technical data; articles containing matter that is published simultaneously in ten or more other journals; material of merely local interest, e.g. "Gravel deposits in Scotland," "Properties of soils in the Lake Ontario region," and so on. But not only are articles often met within the capitalistic journals of little real content or of propaganda and publicity type; we often find also those that are of pseudo-scientific nature, tending to deceive or disorient the reader. Some of this material which, at a preliminary glance, may seem useful, on further examination has to be rejected as of no interest for information purposes. (258)

Journals can be classified in a number of different ways:

1. By subject, whether they are general journals, like *Nature* and *Science,* or highly specialized
2. By format, whether they are devoted to original research reports, reviews, abstracts, news, or any mixture of these

3. By readership, whether they are directed to the research investigator, the clinical specialist, the general practitioner, students, lay readers, or any other specialized group
4. By geography, whether they represent the interests, points of view, or output of a particular area or institution
5. By sponsorship, whether they are supported by a professional society, institution, or commercial publisher (379).

Journals are also classified according to quality, a judgment any indexing or abstracting service, or indeed any library, makes when it selects those to include in its source material. The National Library of Medicine makes use of subject specialists and other consultants in selecting titles to cover in *Index Medicus* and its other databases. Inclusion in any one of the standard indexing and abstracting media is eagerly sought after by editors of new journals, because in some cases inclusion may mean the survival of their journal. Quality is to some extent an elusive measure influenced by changes in editors and editorial policy and changes in taste. One reviewer used language which is reminiscent of the wine connoisseur rather than the detached scientist in describing new journals as "pretentious," "stodgy," and "classical," and marking one new journal's appearance as a "well worn performance" (47:497).

Another classification attempts to place journals in general categories depending on their subject orientation:

1. Substance specialty journals, e.g., *Journal of Lipid Research, Nucleic Acids Research.*
2. Subspecialty journals, based on a central theme or phenomenon, class of organism, a method, or an applied field, e.g., *Journal of Immunology, Insect Biochemistry.*
3. Mission-oriented journals which focus on a social or medical problem, e.g., *Cancer Research, Biology of Reproduction.*
4. Multiscience (general) journals, e.g., *Nature, Science.*
5. Hybrid journals dealing with two or more disciplines, e.g., *Biochemical Genetics, Biochemical Pharmacology* (419).

Journals may also focus on certain species, e.g., *Snake, Birds, Man;* on specific organs or organ systems, e.g., *Spine, Chest, Heart, Brain, Circulation;* on physiological process, e.g., *Digestion, Sleep;* on therapeutic approach, e.g., *Chemotherapy;* on disease, e.g., *Diabetes, Cancer;* on discipline, e.g., *Biochemistry;* or on regional emphasis or sponsorship, e.g., *American Journal of Physiology, Journal of Physiology* (British), but *British Journal of Plastic Surgery, Plastic and Reconstructive Surgery* (American). Whether a journal is designated as American or British depends to some extent on who got there first. Many of the titles cited above have been chosen as a recognition of the editor's penchant for brevity. The science historian George Sarton, founder of the history of science journal *Isis,* was asked why he chose this particular title. "Because it is short," he responded, but obviously with tongue-in-cheek, because it also has something to do with the healing powers of the ancient Egyptian goddess.

Journals can also be classified according to other characteristics such as speed of publication, editorial emphasis, and circulation policies. Although all scientific journals presumably try to keep the lag time between receipt and publication of an article to a minimum, there are a growing number of journals which are specifically designed for "rapid communication." *Biochemical and Biophysical Research Communications* was inaugurated in 1959 for the rapid dissemination of information in all fields of experimental biology. It received its first article on June 12 and published it in its July issue and has attempted to maintain this schedule over the years. The journal is composed from copy submitted by the author. The extent to which it has grown is shown by the fact that the first volume, published July to December 1959, contained 362 pages, while the 106th volume, which covers the 2-month period May–June 1982, contains 1,489 pages. Another example is *Life Sciences: the International Medium for Rapid Communication in the Life Sciences* which began in 1962 and is now published weekly. It is also printed in "camera ready" format, that is, from typescripts submitted by the author. Other alternative methods of rapid publishing have also been suggested and tried. One is the so-called "synopsis journal" in which only a brief summary of one or two pages of the original article along with selected tables and references is published. The full paper is made available in different ways, by microfiche issued with the journal, in "miniprint" editions which can rarely be read with the unaided eye, or on demand from a central depository.

Editors of new journals may not always be candid about their motivations for starting new journals, but they usually seem to be starting from a position of guilt, because they frequently start with an apologia for adding one more journal to an already heavy burden of scientific communication. John Shaw Billings put it succinctly in 1879: "It is as useless to advise a man not to start a new journal as it is to advise him not to commit suicide" (41:91). There do not seem to be any studies of the social dynamics of the generation of new journals or of their life and death cycles. Science editors sometimes rationalize the appearance of a new title by referring to the backlog of papers waiting publication in established journals, but most frequently the reason is based on a presumed inadequacy of attention in the literature to a certain specialty or sub-specialty. Often, the reason given is the interdisciplinary nature of the research in a given field, which causes the research literature to be scattered among many journals. For example, *Cell Motility* was inaugurated in 1981 to provide a vehicle for the thousands of papers beginning to appear on the study of cell movements (221).

New journals are also associated with the emergence of new technologies, especially when there are strong commercial interests involved. One example is genetic engineering, which in 1982 already formed the basis for more than 200 companies. The field also spawned a number of new publications which one reviewer classified into "true magazines," "abstract and 'intelligence' networks," and "gossip sheets" (104). Some of these represent a class of journals sometimes called "controlled circulation" publications, because they are usually distributed free to a selected clientele, such as users of laboratory equipment or media, or to particular groups of practitioners. They are entirely dependent either on advertising income or on the support of such groups as pharmaceutical companies. Other publications in this group are represented by newsletters and bulletins which on the contrary are sometimes made available only at high cost because industry and other specialized groups are willing to pay the prices they demand.

Some new journals are associated with the formation of new societies which are also stimulated by the same need for interaction with peers. A recent example is the International Society of Cerebral Circulation and Metabolism formed in 1980. It was preceded by a series of international meetings which began in 1965 as a result of "the fact that a new technology had become available for measurement of regional cerebral blood flow in man and experimental animals" (405). As described by the editor in the first issue of the society's journal, *Journal of Cerebral Blood Flow and Metabolism,* the formation of the society and the journal occurred almost simultaneously. A small group convened at one of the international meetings to discuss the establishment of a new journal. A publisher was approached, but it was soon realized that some kind of formal organization was necessary to monitor the journal, and so it was decided to form the society at the same time.

This sequence is not followed in every case, and sometimes the journal precedes or follows by several years the organization of a society. The *Journal of Biological Chemistry* thus was established in 1905 as part of an effort to establish biochemistry as a separate discipline, but the American Society of Biological Chemists did not come into existence until 2 years later (256). The *Biochemical Journal* started publication in Great Britain in 1906, but it was 5 years later that the Biochemical Society was founded. On the other hand, the Environmental Mutagen Society was founded in 1969, but its journal *Environmental Mutagenesis* did not appear until 10 years later.

As readers find fewer and fewer relevant papers in the more generalized journals, there is a tendency to establish more specialized journals. This reduces the size of the audience for any particular journal and thus raises the cost of individual subscriptions. In this progressive process areas of coverage have become more and more circumscribed. It has been sustained in the past by the fact that the total audience has increased over the years, but it is questionable whether the process can continue indefinitely.

Monographs

Monographs constitute a class only in that they are more or less distinguishable from journals. The term is used largely to describe those publications which are not classified as journals, and as a synonym for the word "book." It is sometimes also used in a more restrictive sense to characterize a comprehensive work on a specific subject. It is in a sense an arbitrary division in the literature created as much by the need to organize materials for use as by the inherent nature of the materials themselves. The units (papers) in a journal are tied together through the name (title) of the vehicle in which they are distributed and through the more or less orderly system of numbering that is used. Monographs are complete in themselves except that they too may appear in numerous volumes. These distinctions were not as clearly made in the early history of scientific publishing. Books and monographs, like encyclopedias, were often sold in parts and sometimes by subscription, much as Charles Dickens' novels were distributed later in the nineteenth century. There were also journals which, for want of contributors or because of the editor's predilections, were essentially single-author publications. There is a difference between monographs and journals in that, once committed to a journal or series title, individuals tend to continue to receive the successive issues indefinitely. Publishers like to package monographs in series to appeal to the collector's instinct which many people, particularly librarians, have difficulty resisting and also because it may make promotion and sales easier. Many books in which each chapter has an individual author are not much different in content than a journal issue. The sharp division made in modern times between monographs and journals has in some ways created an artificial dichotomy in which one must go to one location to gain access to books and to another to gain access to journal articles.

Monographs cover as wide a range of types and classifications as journals. They include anything from short textbooks which are basic introductions to a subject to the most advanced discussions of highly specialized subject areas. They can be single- or multi-authored works, can be the results of symposia or conferences, and can appear in single or multiple volumes. Textbooks can also vary widely in scope and complexity. Although each one can be regarded as a compilation which is considered fundamental in a particular subject area, they reflect widely divergent views about what information is essential and how it should be organized, and the different pedagogic philosophies of the author or authors. A textbook in microbiology which focuses strictly on the morphology of microscopic organisms does not help much when one needs to learn something about the laboratory diagnosis and disease processes to which they may be related. It should also be remembered that textbooks go out of date rapidly. The life cycle of a typical textbook is regarded as 5 years, which is a short period considering that 2 of those years may be spent in the process of preparing and printing the manuscript.

There are other ambiguous terms used to designate information sources such as handbook, manual, and encyclopedia which can be considered along with monographs because they impinge on one another and also are accessed in a different way than journals. There are actually books designated as handbooks which can be held in your hand or slipped into your pocket. In practice, however, the term has been used for anything from such slender volumes as these to compendia in a hundred or more volumes, like the *Handbuch der biologischen Arbeitsmethoden,* a comprehensive work on biological and biochemical methods, clinical diagnosis, and pathology, edited by Emil Abderhalden and published in Berlin from 1920 to 1939 in 107 volumes. This is an example of a number of "handbooks" issued in Germany in the last half of the nineteenth century and later, some of which are still continued today. They are like encyclopedias in that they attempt to summarize the state of knowledge at a particular time, but resemble journals or books-in-parts in that sections are issued separately in bound or unbound form, sometimes out of numerical order when a subject assigned a later place in the outline of topics is completed and published before an earlier one. Another complication occurs when earlier volumes or sections are published in new and updated versions before the entire work is completed.

Some of these handbooks provided comprehensive and authoritative surveys of special subjects by leading exponents in the field in their day. The format still continues in use today with such titles as *Handbuch der mikroskopischen Anatomie der Menschen* edited from 1927 by Wilhelm von Möllendorf. An example of the type of material it contains is volume 7, part 6, which was published in English in 1979 with the title *Prostate Gland and Seminal Vesicles* by Gerhard Aumüller in 380 pages with 1,625 references. Another example is the *Handbuch der experimentellen Pharmakologie,* which was first issued by Arthur Heffter in 1920 and later by Wolfgang Heubner. It now bears the title *Handbook of Experimental Pharmacology,* and the sections, which could just as easily be called monographs or reviews, appear largely in English. An example is *The Anatomy, Chemistry and Physiology of Adrenocortical Tissue* by Helen Wendler Deane, published in 1962 in 185 pages with about 1,000 references.

The format has also been adopted by the American Physiological Society for their *Handbook of Physiology* published from 1959 to 1977 in nine sections and twenty-six volumes. In 1977, they started the cycle all over again with new articles on the same or revised topics. The rationale for the series was clearly stated by the Board of Publications Trustees of the Society in the foreword to the first volume:

> The original literature in the field of physiology has become so vast and is growing so rapidly that the retrieval, correlation and evaluation of knowledge has become with each passing year a more complex and pressing problem. Compounding the difficulties has been the inevitable trend toward fragmentation into smaller and smaller compartments, both of knowledge and research skills. This trend is not only inevitable, but it is necessary to healthy growth. It must, however, be accompanied by the development of mechanisms for convenient and reliable reintegration in order that knowledge shall not be lost and research effort wasted. (204)

When one looks at the contributions of the individual authors, it is again difficult to distinguish them from reviews or, in the case of longer contributions, from monographs. One chapter in volume 7 (*Parathyroid Gland*) of section 7 (*Endocrinology*) is entitled "Bone structure and mechanism of calcification" and is ninety-two pages long and contains 371 references. Some of these multi-authored works are "analyzed in library catalogs" (parts entered as separate works), but many fall outside the mainstream of standard access tools and must be approached through personal knowledge of their existence. Some of them may also appear in periodical indexes and indexes to reviews, which are discussed in the next chapter.

Chapter **4**

Varieties of Information Sources: Secondary

Reviews

The difficulty of dividing the scientific literature into primary (original) and secondary (derivative) sources has been apparent in the discussion of journals and monographs. In a sense portions of all scientific literature are secondary in that they are based on previously reported work. This is demonstrated by the literature review which has traditionally been a part of every research article, and in which the author gives credit to his predecessors and documents the procedures and findings which have been previously reported. As the volume of literature has grown, editors have tended more and more to restrict literature reviews in research papers to those articles of most direct relevance to the work reported. Accordingly, there is a great need today for publications and articles which are devoted exclusively to reviewing the literature.

Reviews can be regarded in some ways as an interim form between primary and secondary publications. They are secondary in that they are drawn from the primary literature, but primary in that they can present creative new syntheses and, through evaluation and commentary, provide an important means for arriving at scientific consensus. A good review has been called:

> . . . a virtuoso performance in creative communication—retrieval, evaluation, analysis, synthesis, exposition. It may yield information where before there were only facts. It may suggest theories, where before there were only viewpoints. (171:3)

Reviews generally rank high in user surveys as sources of scientific information. Consulting reviews is one of the first steps an investigator takes in approaching a new problem in science or becoming up to date in any subject. From the start, the investigator wants assurance that no one has undertaken or solved the problem previously. Secondly, it is important to know what related work on the problem has been published. A review thus provides an excellent place to start a literature search since, if it has been prepared conscientiously, it eliminates the need to search the literature for the period it covers. It also assists in keeping up with the developments in any particular field by presenting the relevant literature in a systematic and logical narrative. In many fields there are review media which systematically cover the literature at frequent intervals, often annually.

The scope of reviews varies from very broad to very narrow subject areas, and levels of completeness differ depending on the reviewer's orientation and objectives. Reviews can have only a few references if the topic is new or the literature is sparse, but hundreds and even thousands of references may be included in reviews covering well-established fields of inquiry. If the *Guinness Book of Records* had an entry for the number of references in a review article, it would probably cite Dr. Harry Goldblatt. Dr. Goldblatt's major contribution to medicine was a delineation of the role of decreased blood supply to the kidney in the pathogenesis of hypertension. The review to which I refer, however, was one on ricketts—a subject which engaged his attention while he was working in England and began to associate the occurrence of ricketts with the lack of sunshine. His review "Die neuere Richtung der experimentellen Rachitisforschung" was published in 1931 in one of the established review media in Germany called *Ergebnisse der allgemeinen Pathologie und pathologische Anatomie.* It contains 2,723 references, beginning with Glisson, who first described the disease in 1650 (186). The *Ergebnisse* (results) was one of a large number of review media published in Germany before World War II, some of which had their origins in the middle of the nineteenth century. They had titles like the foregoing *Ergebnisse,* or *Jahresberichte* (annual reports), or *Fortschritte* (advances) and were published in a large number of disciplines and subjects. These terms all have their cognates in English and other languages, such as the *Annual Review* series issued by Annual Reviews Inc., a publishing company which began with the *Annual Review of Medicine* in 1950 and by 1981 had reached a total of twenty-four series when it introduced its *Annual Review of Nutrition.*

Reviews have been classified into those which are discipline oriented and which attempt to provide a descriptive review of the annual contribution in a discipline or sub-discipline (e.g., *Annual Review of Entomology*), and those which are categorically oriented and which focus on a scientific problem which engages the attentions of workers in several disciplines (e.g., *Advances in Cancer Research*). The function of the first is to provide an annual record, which is usually comprehensive and noncritical. The second tends to be more selective, critical, and heuristic and frequently indicates the direction of research in that field (7).

In addition, reviews can be classified according to the publishing sources or frequency of publication. For example:

- Reviews are published in so-called primary journals, whose essential mission is to publish original papers.
- Conference proceedings, which can appear in any of the formats cited before, frequently include review papers.

- It has been estimated that there are perhaps 10,000 chapters a year in multi-authored works, treatises, series, and encyclopedias such as the *Handbooks* discussed earlier which can be classified as reviews.
- Review series are published annually or sometimes at greater intervals, e.g., *Annual Review of Microbiology, Advances in Protein Chemistry.*
- Review journals are also published quarterly, e.g., *Biological Reviews, Physiological Reviews* (166).

Reviews in primary journals account for a large part of all published reviews. A 1964 study reported that 67 percent of the reviews in that year appeared in that format. A study in 1972 found that, of 22,000 reviews in all fields, 18 percent appeared in conference proceedings of various kinds (474).

It is fortunate that so many reviews appear as journal articles, because then they are picked up in the standard indexing and abstracting media such as *Biological Abstracts, Chemical Abstracts, Index Medicus,* and *Science Citation Index. Biological Abstracts* groups all the articles which have the word review in the title under that word in its permuted subject index. The first abstracts under any subject in the weekly issues of *Chemical Abstracts* are generally review articles if any are included. *Index Medicus* provides extensive coverage of the various review media in the life sciences, although some, such as *Advances in Lipid Research* and *Biological Reviews,* are covered selectively, depending on whether the articles are regarded as being within its scope. From 1956 to 1966 the National Library of Medicine published a separate *Bibliography of Medical Reviews* which included all the articles identified as reviews for *Index Medicus.* The first six volumes, 1955–1961, were also issued as a cumulated volume in 1961. In 1967, this publication was consolidated with *Index Medicus,* published in the front of each monthly issue, and cumulated annually as a special section of the *Cumulated Index Medicus* with the same subject–author approach as the rest of the index. For about 20 years the National Library of Medicine has also been publishing a series of reviews or bibliographies on separate subjects under the title *National Library of Medicine Literature Searches* which are generated by requests made to its reference staff. Selected searches, which are announced in the *NLM Newsletter* and *NLM Technical Bulletin* and listed in the prefatory pages of the monthly issues of *Index Medicus,* are made available to the public. About twenty-five or thirty are announced every year. Number 82-8 of this series (i.e., the eighth issue in 1982), which is entitled "Aquatic toxicology of metals and metallic compounds," contains references, and covers the literature from January 1972 to May 1982. The *Science Citation Index* codes all review articles with an "R." Since 1974, the Institute for Scientific Information has also been publishing the *Index to Scientific Reviews,* which covers many of the multi-authored works as well as the articles identified in its screening process for review articles for the *Science Citation Index.* The formats for both publications are similar in including permuted term indexes, source, citation, and author indexes. (The titles and sources of reviews cited here are only a few of many which are available and which can be identified in the literature guides and bibliographies listed in the Appendix.)

New and innovative approaches to reviews, using the new computer-based technologies, are also being investigated. One such effort is the "Hepatitis Knowledge Base" being explored by the National Library of Medicine. From 20,000 papers on the subject, forty recent reviews were selected out of which a base of 575 articles was developed for review by a group of experts. These were then reduced to an organized body of knowledge, structured under topic headings. The "knowledge base" (review) was produced by a novel method called "computer conferencing" in which the geographically dispersed experts could enter information from their own terminals (38). Another such effort is the Institute for Scientific Information's *Atlas of Science,* which published its first volume in 1981 under the title *Biochemistry and Molecular Biology* (229). The "atlas" uses a technique called "cluster mapping" in which computer-based citation analysis is used to show the relationship of a group of research papers which cite each other and thus tend to define an area of research interest, or a "research front" as the Institute calls it in another application. The papers which made the most important contribution to the subject are mapped, that is, shown in a graphic representation of the connections between the papers. The first volume covered 102 such "research front specialties" including such subjects as "DNA replications of proteins," and "Insulin binding in obesity." In addition to the maps, each subject includes a "minireview" which provides a history of the topic and a summary of the recent work. The most frequently cited and citing documents are also included.

Indexes and Abstracts

I am discussing indexing and abstracting media together here because the lines between them tend to become blurred with increased use of computer-based and online systems. Many of the abstract publications such as *Biological Abstracts, Chemical Abstracts,* and *Excerpta Medica* include only title information for some of their entries. On the other hand, online systems such as MEDLINE, a computerized version of *Index Medicus,* is said to have abstracts available online for as many as 40 percent of its entries.

Index and abstract publications have been produced in large numbers covering many general and special fields in the life sciences. In science and technology there are about 1,000 abstracting and indexing services in the world, with 330 of them in the United States alone (246:149). These services relate primarily to the journal literature, although book catalogs of special library collections could be included here as well. In a sense, the card catalog of any library is an index which is unique for a special collection. Each indexing and abstracting service has its own scope, coverage, access mechanisms, and methods of vocabulary control. A user should become familiar with these aspects before consulting any of them. They are, moreover, constantly changing if they are issued serially. All indexing and abstracting services vary in coverage of the

literature; some try to be comprehensive, and others are highly selective. There
is also a considerable overlap among some of them.

The American National Standards Institute defines an abstract as

> . . . an abbreviated, accurate representation of the contents of a document, pref-
> erably prepared by the author(s) for publication with it. Such abstracts are also
> useful in access publications and machine-readable data bases. (16)

By calling attention to author-produced abstracts the definition underlines the
fact that more and more journal editors are requiring authors to submit abstracts
of their articles for publication along with the article. Writing good abstracts,
therefore, is a necessary skill, particularly since author-written abstracts also
serve as the input for abstract publications. Abstracts are also produced by a
vast army of paid and voluntary abstractors scattered all over the world who
are generally subject specialists. Abstracts vary widely in purpose and comprehensiveness. Their length frequently depends on the nature of the article. The
shortest example is a one word abstract of a discussion of the correct spelling
of the name of the originator of the perforated-plate funnel used by organic
chemists. The title of the article was "Büchner or Buckner?" and the abstract
was "Büchner" (49:11).

Abstracts have been divided into three categories. The *indicative abstract*
may contain information on the purpose, scope, and methodology of the research
reported, but does not include information on results, conclusions, or recommendations. It serves to supplement the information supplied by the title of the
article and helps the reader to decide whether to consult the original article.
Informative abstracts provide all the information included in the indicative abstract as well as some of the qualitative and quantitative information in the
original, including ranges of data, formulas, and findings. The term *critical
abstract* has been used for those in which the abstractor compares the data in
the original with other reports or provides other evaluative remarks (95).

Indexes and abstracts can be used in various ways. They can be used as
current awareness tools by systematically screening each issue of the publication
as it appears, under the pertinent terms in indexes or in the classified sections
of the abstract journals. They can also serve as more efficient screening devices
in searching the literature, since they assist the searcher in deciding whether
the original article is relevant and should be consulted. Sometimes informative
abstracts supply all the information required and eliminate the need to consult
the original article. Abstracts also provide access to the literature in languages
in which the reader is not fluent or knowledgeable. Abstracts in English are
frequently supplied with original articles published in other languages. This
service can circumvent or eliminate the expensive process of translating an article
which seems pertinent but may not prove to be so.

In the past few years, indexes and abstracts have been more and more made
available for searching on computer terminals, as online databases. These include
Agricola (the online name for *Bibliography of Agriculture*), MEDLINE *(Index*

Medicus), PsychINFO *(Psychological Abstracts),* BIOSIS *(Biological Abstracts),* CASEARCH *(Chemical Abstracts),* and SCISEARCH *(Science Citation Index).* In many ways these online databases have facilitated searching the literature, by making it possible to search large amounts of literature in a short time and introducing refined search techniques.

Data Collections, Dictionaries, Manuals, Etc.

As we have seen from my earlier discussion, there are many terms such as "handbook," "encyclopedia," and "dictionary" which do not help much in defining the format in which information is presented. Along with the term "manual," they can be used for anything from a slender volume to a huge compendium and may refer to works in tabular, narrative, or any other form. A bibliography of handbooks and tables in science and technology published in 1979 contained 2,027 entries (362). These publications are sometimes also called "reference works" because we do not generally read them cover to cover. Many handbooks will not fit anyone's hand, and some encyclopedias are not very encyclopedic. The *McGraw-Hill Encyclopedia of Science and Technology,* for example, has fifteen volumes in the fifth edition (New York, 1982) and covers a broad range of subjects from aardvark to a family of fungi called zythiciae. The fifth edition of *Van Nostrand's Scientific Encyclopedia* has two volumes containing 3,067 pages (New York, 1982) and is similar in scope. The articles in both are arranged alphabetically, but they differ in length and in the topics or terms selected for discussion. The *Van Nostrand,* for instance, has an article on "abortion" but not on "abnormal behavior," while the reverse is true of the *McGraw-Hill.* It would, in fact, be an interesting exercise to analyze how each of these encyclopedias chooses to anatomize and organize the whole body of scientific knowledge.

There are also "encyclopedias" in many special subjects such as the *Encyclopedia of the Biological Sciences,* edited by Peter Gray (2nd ed. New York: Van Nostrand Reinhold, 1970). It provides discussions of biological concepts like conjugation and reproduction, defines and describes biological species, and includes short biographical sketches of outstanding biologists. Except for the length of some of its entries, it might just as easily be described as a dictionary, just as some "dictionaries" go beyond the bare definitions of concepts to provide information that might be regarded as encyclopedic.

Dictionaries as a group tend to fall into two general classes. In the first the concepts are defined in the same language, covering the language in general or the language for a specific discipline. One example is *Henderson's Dictionary of Biological Terms,* which has appeared in nine editions at intervals of 3 to 15 years (9th edition prepared by Sandra Holmes, New York: Van Nostrand Reinhold, 1979). Dictionaries are important because communication takes place only when two individuals share the same concept with the same name of that concept. Control of the terms and consensus about their meanings, therefore, has been an important role played by societies in various disciplines. Regulatory groups, particularly in the taxonomic sciences, and frequently on an international level, exercise rigid control over their terminologies. In anatomy, for instance, there have been three major changes in codes of nomenclature. These are all reflected in the *Anatomical Dictionary with Nomenclature and Explanatory Notes* by Tibor Donath (Oxford: Pergamon Press, 1969). It compares the changes in the Basle Nomina Anatomica (1895), the Jena Nomenclature (1935), and the most recent international revision, the Paris Nomina Anatomica (1955).

Another major grouping of dictionaries is those in which concepts in one language are expressed in their cognates or comparable terms in another language. These also can be general like *Cassell's German-English, English-German Dictionary* (New York: Macmillan, 1978), or highly specialized like the *Medical and Pharmaceutical Dictionary: English-German* by Werner E. Bunjes (4th ed. New York: George Thieme Verlag, 1981) and his *Worterbuch der Medizin und Pharmazeutik, Deutsch-Englisch* (3rd ed. New York: George Thieme Verlag, 1981).

There are also compilations which collect data of various kinds from a large number of sources and bring them together in a format in which they can easily be consulted. These publications are sometimes called tertiary instead of secondary publications, because secondary sources are used to locate the primary sources in which the data are first reported. Since science is largely a quantitative discipline, numerical values relating to many phenomena in nature have been collected over a long time. These data compilations are frequently presented in tabular form, but they may also be in narrative form in subjects in which the data are qualitative, such as the representation of genes in the chromosomes. Biological data are said to differ significantly from data in chemistry and physics

> ... because the objects of the observations and measurement are living organisms. In the laboratory, the variability *between* subjects and the variability *within* subjects precludes the determination of absolute values, such as are available in chemistry and physics, but nonetheless provides a statistically meaningful measure. (240)

The *International Critical Tables of Numerical Data, Physics, Chemistry, Technology* published in seven volumes under the auspices of the National Research Council between 1926 and 1933 is still useful. A more familiar compilation of similar data frequently found near the laboratory bench and in most scientific libraries is the *Handbook of Chemistry and Physics,* which has been published over 69 years. The 63rd edition appeared in 1982 with almost 2,500 pages and a comprehensive index (Boca Raton, Fla.: CRC Press, 1982). The same company has issued a whole series of "handbooks" in several disciplines including a *Handbook of Microbiology* in four volumes in 1974 and a *Handbook of Biochemistry or Molecular Biology* in 1976. In 1971 they published a composite index to ten handbooks of this kind to help the user know to which handbook to turn for particular data, and many of these have since been revised.

Data relating to drugs lend themselves particularly well to these kinds of compilations, such as official pharmacopoeias and formularies which have an ancient lineage. The *Merck Index; an Encyclopedia of Chemicals and Drugs* has been published since 1889. The tenth edition, which appeared in 1983 (Rahway, N.J.: Merck and Co., Inc., 1983), contained 10,000 "monographs" which ranged in length from a few lines to half a page or more, representing the most important chemicals, drugs, pesticides, and biologically active substances. A recent survey (399) listed thirty-three drug compilations which are currently being published in English, and other countries provide counterparts in many other languages. The list of thirty-three represents only a small number of the total drug information sources available. The survey lists fourteen different kinds of data which can be supplied by these compilations. None of the thirty-three titles supplied all of them. Only eight provided descriptions of the chemical and physical properties of the drug, only eleven gave information on toxicology, and five did not even supply information on therapeutic use, but served only to supply information for identification and determination of market availability (399:6).

A source for such information as the dimensions of the eye of a gnat or the thickness of an Australian aborigine's hair is an important series of data compilations published sporadically since 1925, first from Berlin and then from the Hague, under the title *Tabulae Biologicae.* It was resumed briefly after World War II but now no longer seems to be published. Volume 20 of the series, which appeared in 1941 in 963 pages, bears the title *Growth of Man* and has been called the standard work for anthropometry of all the world. It was edited by the well-known American anthropologist Wilton Marion Krogman. The *Tabulae Biologicae* have been replaced but not completely supplanted by such publications as the *Biology Data Book,* compiled and edited in a second edition by Philip L. Altman and Dorothy S. Ditmar, and published by the Federation of American Societies for Experimental Biology in three volumes in 1972. It supersedes some of the data collections published earlier by the Federation, such as those on *Blood and Other Body Fluids, Metabolism, Respiration,* and *Circulation,* but not the one on *Growth.*

Another useful compilation of biological data, this time specifically about the human species, is the *Report of the Task Group on Reference Man* issued by the International Commission on Radiological Protection (New York: Pergamon Press, 1977). The purpose of the report, as defined in the introduction, is the "estimation of radiation doses to the human body, whether from external or internal sources . . ." (227:1). In the process all kinds of human biological data are tabulated, including, to stay on a familiar subject, the weight of human hair, while at the same time admitting that some factors are difficult to calibrate:

> Obviously the weight of hair on the head is partly a matter of the individual's endowment and if suitably endowed—may vary with the trends or fashion in taste. (227:58)

Although many data retain their validity for a long time, data collections frequently go out of date. The turnover in drugs, with new ones being introduced every day and old ones going quickly out of fashion, is particularly high. New data in all fields are constantly being added to the storehouse of knowledge. Reviews and other publications must frequently be consulted to acquire the most recent data. For this reason and also because numerical and tabular data lend themselves well to this kind of handling, many databases of this kind are being made available through computer terminals, often from distant computer storage repositories. Chemical and pharmaceutical data are now well served through such channels. These databases are sometimes called non-bibliographic databases because they supply information directly rather than citations to publications. One such database is called CHEMLINE. It is produced by the National Library of Medicine (NLM) Toxicology Information Program and contains the nomenclature and index of chemical structural fragments for over half a million compounds. It also serves as an adjunct to another NLM database called TOXLINE (Toxicology Online), which contains over 1,250,000 items and is adding data from over 100,000 animal and human toxicity studies a year. These are only two of a large number of bibliographic and non-bibliographic databases which are being made available for online searching. There is even one called HORSE produced by Bloodstock Research Information Services, which supplies data on pedigree, breeding records, and racing performance for all thoroughbreds in North America and should prove useful to all racetrack devotees (99:100).

Recently, the American Medical Association entered into a joint venture with a communication company which furnishes a network that links terminals to computers and computers to other computers all over the United States. Initially the A.M.A. is making available online several of its standard printed compilations. One is *A.M.A. Drug Evaluations,* now in its fifth edition. It is published in conjunction with the American Society for Clinical Pharmacology and Therapeutics and provides information on dosages, adverse reactions, toxicity, and literature references on drugs in current use. The online version will be updated each month and provide additional information as well as greatly enhanced searching capabilities. Other publications being made available online are A.M.A.'s *Current Medical Information and Terminology,* which provides descriptions of diseases, disorders, and conditions, and its *Physicians' Current Procedural Terminology* used in systems of medical reporting, which will also be updated more frequently than their printed versions. Other databases being projected include laboratory data, adverse drug reaction reporting systems, and poison control information, as well as a number of other services including "electronic mail" through which any subscriber to the system can send a message to any other subscriber through his or her computer terminal.

Report Literature and Dissertations

There is another publication class or format which, although not as significant in the life sciences as it is in physics and engineering, can still serve as a useful information source. It is referred to as the technical report literature, or sometimes as "unpublished reports," because they are usually produced in association with contracts in research and development projects, or as interim or final reports for research grants. They are distributed selectively and therefore not regarded as part of the "open literature." They are, nevertheless, produced in large numbers, and since many of them do not end up as articles in standard scientific journals, they must be sought out through different channels. The National Technical Information Service (NTIS, a branch of the U.S. Department of Commerce) is charged with collecting and processing technical, scientific, and engineering reports from both domestic and foreign sources. Most of the reports are derived from work performed in or subsidized by federal agencies, but some reports from state, non-governmental, and foreign agencies are also included. There are over 1 million reports on file which can be made available in full form or microforms. Every 2 weeks NTIS publishes a voluminous index to the reports added to its files. Called *Government Reports Announcement and Index,* it is divided into twenty-two categories, six of which specifically relate to biological or medical sciences. Each issue includes a type of subject index called "keyword out of context" as well as personal, author, corporate author, and other indexes. NTIS also supplies a number of other services such as a weekly newsletter outlining new reports in a number of subject fields, as well as "published searches" which it has prepared in over 4,000 subject areas. Its database is also available for searching online through some of the same channels as the other online databases I have cited.

Academic dissertations can be considered along with technical reports, because they also do not constitute a part of the "open literature," have limited distribution, and must be accessed in different ways. Dissertations are part of an ancient tradition going back to the medieval university where, however, they played quite a different role than they do today. In the past they served the candidate as a means of demonstrating his knowledge of the learning of the past, rather than as means of adding to the corpus of knowledge. The professor served as *praeses* and presided over the presentation of a thesis which the candidate or *respondent* had to defend in public. These dissertations were printed with the names of both the *praeses* and *respondent.* With some dissertations published as late as the eighteenth century it is not clear whether authorship should be attributed to the *praeses* or *respondent.* Thus, most of the dissertations of the students of Linnaeus which were published from 1749 to 1770 in Stockholm, in a seven-volume collection called *Amoenitates Academicae,* are said to be the work of the great naturalist himself (231). It has also been suggested that some professors may have assumed this prerogative even later than the nineteenth century, by assigning their names to research papers which came from their laboratories.

Dissertations in the modern university have assumed quite a different role: they now serve to demonstrate the candidate's ability to perform independent research and to add to the store of knowledge. They can also serve other useful purposes. Since the candidate generally carries out a comprehensive survey of the literature in conjunction with a research topic, the published dissertation can sometimes serve the purpose of a literature review in that subject area. As graduate programs have grown, so has the number of dissertations. In the United States they increased from 21,775 in 1968 to 32,705 in 1977 of which around 20 percent in each of these years fell into subject areas related to the life sciences (425:78). There seems to be some variation between disciplines, but it was found that dissertations generally result in at least one publication in the open literature (361).

However, one need not wait until dissertations are published, usually in abbreviated form, in journals. More than 430 academic institutions in the United States and Canada deposit copies of dissertations submitted by their students with an organization in Ann Arbor, Michigan, called University Microfilms International. Since 1938, it has published a guide to dissertations, now called *Dissertation Abstracts International* since it also includes some foreign universities. It is published in two series, one covering Humanities and the other covering Science and Engineering. Each issue includes author indexes and keyword indexes to the titles which are cumulated annually. The database is now also available online as the *Comprehensive Dissertation Index,* which indexes all dissertation titles in the University Microfilms file since 1861.

Research in Progress

I could just as easily have started this survey of information sources in the life sciences by considering sources of information about research in progress rather than conferences and journals, since research initiates the process which results in the report included in these forms. An investigator can determine whether anyone has worked on a problem by determining if it has been reported in the literature. The literature search, however, will not reveal whether someone is currently working on a problem. Some of this information is available to the members of the invisible college, if they know the investigators or they are made known to them by their colleagues and associates. Some research at this stage is reported in the form of preliminary notes in general journals such as *Science* and *Nature.* It may also be presented at conferences and published in abstracts in such places as the *Federation Proceedings* of the Federation of American Societies for Experimental Biology, which twice a year publishes abstracts of the papers which are to be presented at the meetings of its constituent societies. As an example, the abstracts of the papers scheduled for presentation at the 66th Annual Meeting in New Orleans, 15–23 April 1982, were published in three numbers of the *Federation Proceedings,* 1–10 March 1982. They included 8,754 abstracts along with an index to the authors and title words in the papers.

There are also a number of sources which may provide information about research that is in progress and has not yet been reported in the literature in any way. Since not all research in progress is reported, these sources are by nature incomplete. Some are available in published form, but others can only be accessed online through computer terminals. The U.S. Department of Agriculture, for instance, makes information on research in progress available online through a system called CRIS/USDA (Current Research Information System).

It includes notices of research funded by the Department as well as reports of research submitted by cooperating institutions such as colleges of veterinary medicine.

Since 1961, the National Institutes of Health has published every year a compendium called *Research Awards Index,* which lists health-related research currently being conducted by nonfederal institutions and supported by the health-related agencies of the Department of Health and Human Services. Related basic and social science research is also included, as well as notices of publications and travel grants. The grants are listed first under about 7,000 subject headings in alphabetical order, with supplementary indexes which help to identify the principal investigators and the grant or contract number, but very little information beyond the title of the project is supplied. The information is also available as a computer database called CRISP (Composite Retrieval of Information on Scientific Projects), but it has not yet been made available online.

A much broader information source on research in progress is provided by the Scientific Information Exchange (SIE). It was called the Smithsonian Scientific Information Exchange until recently when it was transferred to the National Technical Information Service, where it is maintained as an online database available through some of the same channels as other online databases. The SIE database provides reports of research projects from the current year and two previous years on basic and applied research in the life, physical, social, behavioral, and engineering sciences. The file contains primarily research projects funded by the U.S. government. It includes both intramural and extramural government research, in contrast to the *Grants Awards Index,* which covers only extramural projects. It also includes some research funded by state agencies, foundations, nonprofit institutions, and foreign governments. In April 1981 there were 342,711 research reports on file in this system, and 9,000 new ones were being added each month. The file can be searched by title words and words in the abstracts. It also provides such information as starting and completion dates of the projects, and names and addresses of the principal investigators.

In the last two chapters I have surveyed some of the formats in which information is made available in the life sciences. In their totality they may resemble a marketplace with all kinds of commodities available rather than an organized system, though some elements of organization can be observed. In the circumstances one can easily understand why some workers in the life sciences complain about "information overload," a situation which prevails when any system, human or mechanical, is presented with more information than it can assimilate. In information systems there are two factors which provide some relief: redundancy and obsolescence. As we will see in the next chapter there are some rough measures which have been suggested for degrees of obsolescence as they occur in the literature. We do not appear to have any adequate measures, however, of redundancy, the degree to which information is repeated in one format or source or another. Though this redundancy may seem inefficient and wasteful, it may also be a factor that helps us to cope with the system.

Chapter 5

Characteristics of the Literature

In the past few years there have been so many studies of the characteristics and use of the scientific literature that we are threatened with a new kind of literature glut. How much relevance any of these studies has to the day-to-day activities of the working scientific investigator or practitioner is difficult to know. They do, however, add to our knowledge of the literature as part of the environment in which we are working. Many of these characteristics have been subjected to mathematical manipulations. One problem with studies of such diverse populations as scientific journals and scientific papers is that all the items tend to become equivalent, and the paupers occupy the same space in the distributions as the kings. The numbers depend to a large extent on the time the data were collected, the ability to differentiate and define the units being counted, and the sources used by the compilers. Estimates of the size of the population of scientific–technical journals and their subject and language breakdowns therefore tend to vary considerably.

Size and Growth

Recent estimates of the number of scientific and technical journals in the world today range anywhere from 20,000 to 100,000. The latter figure is apparently based on a projection made by Derek de Solla Price in 1961 to dramatically demonstrate the exponential growth of science since the eighteenth century. Taking a base of ten scientific journals in existence in 1750, he showed that they increased by a factor of ten every 50 years, and this rate projected to the year 1950 indicated that there would at that time be 100,000 (372:97). King, surveying the scene in 1960, estimated that there were 60,000 scientific and technical journals in existence at that time (246). There is some evidence that there were more than fifty journals which could be labeled scientific or technical in 1750 (260:78), but this does not change the fact, as Price indicated later, that many of these titles were and are of short duration. Of the almost 500 substantive scientific and technical titles, for instance, which appeared between 1665 and 1790 only thirty-four had an existence of more than 20 years (260:93).

Projections are in any case notoriously dangerous affairs. As early as 1881 John Shaw Billings, the architect of the National Library of Medicine, saw the absurdity of projecting the increases in numbers of periodicals indefinitely, pointing out that if the ratio of geometric progression continued "our libraries will become large cities" (41:128). There are obviously some limiting factors: journals cease publication and they are subject to such things as changes of names, consolidations with other titles, or splitting into two or more titles. Unfortunately, few data on these factors seem to exist.

In 1963 Gottschalk and Desmond estimated the worldwide distribution of scientific and technical periodicals in the natural, physical, and engineering sciences, experimental psychology, and physical anthropology. They used current national press directories as well as the extensive holdings of the Library of Congress as a basis for their calculations and arrived at 35,000 scientific–technical periodicals in existence in the world at that time (190:188).

The National Lending Library in Great Britain, an agency of the British Library that serves a worldwide clientele, attempts to obtain as comprehensive as possible a collection of scientific and technical periodicals. Its detailed estimate of materials in this class in 1967 was 26,000, based on the total number of titles currently received and on order in 1965 (23). Another study was carried out by the same institution in response to charges of publishers that large-scale photocopying was adversely affecting the sale of their journals. The study provided persuasive evidence that it would be difficult for the publishers to show any financial damage and at the same time indicated that the Library was receiving 44,767 serial titles (277). How much of this increase was a result of change in scope and policy of the Library's collecting activities? There were surely other factors besides the growth of the literature to account for the increase from 18,000 cited for 1964.

Discrepancies in two recent studies of scientific journal publication in the United States underline the difficulties in arriving at accurate estimates. Both studies were concerned with the economics of journal publication. The first, which identified only 2,459 titles, was based on 1973 data on "U.S. scholarly and research journals" including those in the social studies and the humanities (464). The other study, based on 1977 data, divided the journal population into one group which was subject to refereeing or peer review and another which was made up of non-refereed journals such as newsletters and trade publications. By combining various sources along with data derived from their own surveys the study team identified 4,447 titles in the first group and 4,468 in the second (246).

Some evidence exists that the growth in the number of scientific–technical journals has been offset by the number which cease publication. Gottschalk and Desmond cite a study of journals on radioactivity which showed that two-thirds of them had come into existence between 1900 and 1930, and that one-third of the entire list were no longer being published by the mid-1950s. They cite another study of aeronautical journals published during the period 1900 to 1962 which found that only 1,553 out of 4,551 titles were still current in 1962, a mortality rate of 66 percent. The mortality rate for journals with their origins in the years 1950 to 1960 alone was 10 percent. They also sampled pages from the *World List of Scientific Periodicals, 1900–1950* and found that 33 percent were no longer in existence (190). Barr sampled the next edition of the same list and found that only 40 percent of the titles were still current (23). The mortality rates for biomedical journals found by Orr are closer to the first figure. He analyzed the titles in the National Library of Medicine's *Biomedical Serials, 1950–1960* and found that more than a third of the serials which had been alive in 1950 had died by 1961 (343). The data on the mortality rates of journals are more or less confirmed by data on their longevity. The average age of the scientific–technical journals in existence in the United States in 1975 for all fields was 25.4 years, with a high of 31.2 years for the life sciences and a low, for obvious reasons, for the computer sciences of 14.1 years, although even older disciplines like the physical sciences and psychology averaged about 18 years (246:92).

Estimates and predictions of the growth of the scientific literature are sometimes projected in terms of rates. The figure of doubling every 14 to 15 years is frequently encountered with relation to both the literature and the growth of library collections. This translates to a growth rate of 5 percent per year. Very few of the estimates in the recent literature, however, seem to confirm this projection. Fry and White found that the growth rate for scholarly journals from 1969 to 1973 was more like 2 percent when allowances were made for dead journals (143). Orr found that the number of "substantive biomedical serials" had increased by 7.5 percent in 1960 over the number published in 1950, which translated into a rate of doubling every 38 years. When, he asked, does a rate cease to be exponential and become merely arithmetic? (343:1329)

These rates and projections have been correlated to some extent with increases in the size of the population served by scientific–technical journals. In psychology, for instance, it has been reported that the workforce doubled every 10 to 12 years between 1890 and 1959, a rate which has been slowing down to that of doubling every 12 to 15 years. The rates of increase have also been diminishing in other fields according to recent indicators. The increase in science and engineering degrees awarded decreased from an average of 29 percent for the period 1965 to 1970 to 8 percent for the period 1967 to 1972 (247). The relationships between increases in populations of scientists and of journals are somewhat confirmed by the data on the number of articles produced per scientist, which changed little from 0.12 in 1965 (that is, a few more than one in ten published at least one article in that year) to 0.14 in 1977 (246:61).

There is perhaps little comfort or enlightenment to be gained from the foregoing data. Theoretically, the number of new journals which appear every year is balanced to some extent by the number which disappear and is offset by the number of new scientists who come into the marketplace. The difficulties arise out of the fact that scientists are confronted with the same total mass of literature out of which they must select an ever narrower segment with which to be concerned. Numbers of subscriptions to individual journals therefore tend to become smaller and smaller so that a large part of the cost of the system must be borne by institutional subscribers such as libraries, who are required to pay ever higher prices. However, it is not the total mass of scientific journals with which either individual or institutional subscribers are concerned, but only that portion which relates to their particular interests. It is therefore more important to know about the distribution of the journal literature by subject and language.

Distribution by Subject, Country, and Language

The task of determining numbers of scientific journals which fall into different subject areas suffers from the same kind of ambiguities as determining the size and growth of the total journal population. In addition, there are the difficulties of assigning journals to specific subjects. Comparisons between one analysis and another are sometimes difficult because of the differences in the scope of the subject categories used and the lack of agreement about what aspects are included. Many journals are multidisciplinary and cannot be assigned to a single subject. Another problem, as Orr points out, is that the literature can be classified in two ways: in terms of the literature used or generated by workers in that field, and in terms of the actual subject matter of the literature covered. These are not always coextensive (332:1311). An example is provided by an attempt in 1973 to develop a technique for finding all the journals which would be relevant to the development of a toxicology information service. Of a total of 42,240 "primary journal titles" which the investigators screened, they decided that as many as 16,662 could contain material of toxicological–biological interest (122).

The numbers of journals which are assigned to the life sciences, therefore, vary considerably from one study to another. The survey of the life sciences published by the National Academy of Sciences in 1970 indicated that, of 26,000 "distinct scientific journals published annually, the life sciences claim no less than 50 percent (20 percent for agriculture, 13 percent for medical science, 4 percent for basic life sciences, and 10 percent for technology), or 13,000 serial publications" (318:407). The total number of world journals in the life sciences was estimated to be 4,420 excluding agriculture and technology. These figures can be compared with the total number of titles received by the National Library of Medicine in 1978, 18,169, of which 6,019 were published in the United States. Some qualitative judgments can be inferred from the fact that a total of 2,598 of the world titles and only 862 of the U.S. titles were included in *Index Medicus* in that year (89:215). The 1977 survey of U.S. scientific–technical journals cited earlier, on the other hand, identified a total of 1,318 journals in the life sciences which met their peer review criteria (246:83).

Distributions of journals by country are less questionable since they can usually be determined on the basis of place of publication. Comparisons of these

Table 1 Percentage Distribution of Life Science Journals by Country

Country	Source*					
	1	2	3	4	5	6
United States	23.3	33.1	20.1	25.7	17.6	23.5
United Kingdom	7.7	6.9	12.6	4.6	6.2	6.8
U.S.S.R.	7.4	3.5	8.8	4.8	6.2	2.3
Germany	6.7	6.8	8.0	4.6	8.5	8.7
Japan	6.4	7.0	6.1	9.2	7.9	5.2
France	4.1	5.5	5.2	6.0	7.6	8.2
Italy	3.2	4.0	3.4	5.3	4.2	10.0
Other	41.4	33.2	35.8	39.8	41.8	35.3

*1. Biosciences Information Services, 1982, 9,430 titles (42:IV)
 2. Corning, 1978, 18,169 titles in National Library of Medicine (89)
 3. Carpenter, 1973, 24,801 titles in National Lending Library (70)
 4. National Academy of Sciences, 1969, 3,100 Life Sciences titles (318)
 5. Gottschalk, 1961, 35,300 World Science–Technology titles (190)
 6. Brodman, Taine, 1957, 3,597 titles in National Library of Medicine (62)

distributions over a period of time can also be interesting because they provide an indirect measure of the scientific dominance of a particular country. Great Britain, France, and Germany, which are said to have been responsible for 50 percent of all the biomedical journals published in 1875, accounted for less than 20 percent by 1978. Some of the reasons for this phenomenon are obvious, such as the rise of the United States along with the U.S.S.R. and Japan as producers of biomedical literature, as well as the entry of the developing nations into journal publication (89:213). Changes in national dominance are also reflected in the literature of special subjects. One of the earliest studies of the literature as a measure of national dominance was a survey of publications in comparative anatomy from 1543 to 1960 which showed that England, Germany, and France collectively accounted for 70 to 80 percent of the literature during the period (79). The changes from 1907 to 1970 in the origin of the abstracts included in *Chemical Abstracts* provide additional dramatic evidence of this phenomenon. The United States' contribution increased from 3.9 percent in 1907 to 27.4 percent in 1970. The comparable figures for the U.S.S.R. are increases from 5.4 percent to 23.6 percent and for Japan from 0.2 percent to 7.2 percent. In the same period Germany's contribution fell from 52.5 percent to 6.5 percent, the United Kingdom's from 17.1 percent to 6.2 percent, and that of France from 16.8 percent to 4.1 percent. The contribution of all the other countries increased from 4.1 percent in 1907 to 25.0 percent in 1970 (316:81).

I have collated six different analyses of journal lists by country, ranging from Gottschalk and Desmond's survey in 1961 of 35,000 world scientific–technical periodicals to 3,100 life science titles analyzed for the National Academy of Sciences survey published in 1970 (see Table 1). The results show some conformity and some variation based on similarities of populations surveyed and variations in editorial policies of indexing services.

Table 2   Percentage Distribution Scientific–Technical Articles by Language

Language	Source*				
	1	2	3	4	5
English	50.3	75.0	51.2	69.0	62.8
Russian	23.4	10.0	5.6	7.4	20.4
German	6.4	3.0	17.2	7.6	5.0
French	7.3	3.0	8.6	4.0	2.4
Japanese	3.6	1.0	0.9	2.5	4.7
Other	9.0	8.0	16.5	9.5	4.7

*1. *Chemical Abstracts,* 1965, Wood (471)
 2. *Biological Abstracts,* 1965, Wood (471)
 3. *Index Medicus,* 1965, Wood (471)
 4. *Index Medicus,* 1978, Corning (89)
 5. *Chemical Abstracts,* 1978, CAS Today (15)

Distributions by country of origin do not correspond closely to distributions by language, even when we add those for the United States and the United Kingdom to explain the dominance of English. Many journals published outside both these countries are published in English, and authors who use other languages as their native tongue also frequently publish in English. A survey of the 1973 literature in the *Science Citation Index* showed that, of 17,376 articles which had at least one author with a French address, 7,264, or 42 percent, appeared in primarily English language journals. A further analysis confirmed that 61 percent of these had in fact been published in English (158).

The language distribution of the literature is more important to investigators than the distribution by country because it reflects more accurately that portion of the literature which is directly accessible without going through the medium of a translation or an English abstract. By the end of the eighteenth century, Latin had virtually been replaced by a variety of vernaculars as the language of science. For a short time thereafter, French served to some extent as the *lingua franca* of the scholarly world. The Academy of Science established in Berlin in 1700 published its earliest proceedings in French. The Academy which was founded in St. Petersburg in Russia, on the other hand, tended to publish in German because German scientists played a large role in the development of science in that country. It was not, however, until the latter part of the nineteenth century that no scholar with any pretention to thoroughness could avoid learning German. English has largely replaced both French and German as the dominant language of science, but a significant part of the world's literature still appears in other languages, as can be seen in Table 2.

Scatter and Use

There are a number of other factors which tend to reduce the size of the literature with which the investigator and practitioner needs to be concerned.

Table 3   Scatter of Scientific References

Subject	Percentage of Journal Titles Accounting for Each Quarter of the References			
	First	Second	Third	Fourth
Chemistry	0.75	3.05	10.2	86
Physics	0.75	1.5	10.75	87
Biochemistry	0.35	1.05	5.1	93.5
Tissue culture	1.5	7.0	32.5	59

Adapted from Vickery (449), 1938–1953 data.

One is that the relevant sources for any particular subject seem to be concentrated in a relatively small part of the literature. This phenomenon has perversely been called the "law of scattering" and was first described by the British librarian S. C. Bradford, who studied the distribution of references compiled in the Science Library in London in the 1930s. He studied the dispersion of references in a bibliography on geophysics and found that he could put all the journal titles into three groups, each of which contained the same proportion of the total references. The first nine titles contributed 429 references, the next 59 titles contributed 499, and the last 258 journal titles produced the remaining 404. As described in his own words the phenomenon is such that:

> ... if scientific journals are arranged in order of decreasing productivity of articles on a given subject, they may be divided into a nucleus of periodicals more particularly devoted to the subject and several groups or zones containing the same number of articles in the nucleus, when the number of articles in the nucleus will be as $1:n:n^2$. ... (54:116)

A large body of literature has grown up around this "law." It has been modified by introducing other characteristics which influence use such as the age and size of the literature under consideration. Essentially all the studies support the basic principle that a large volume of literature on any subject is concentrated in a relatively small number of journals. Data from studies conducted from 1938 to 1953 in several disciplines showed some variations but similar distributions. The first fourth of the references in these disciplines was concentrated in 0.35 percent to 1.5 percent of the journals (see Table 3). In 1964, the British National Lending Library reported that half of the articles requested for loan were to be found in 7 percent of the 26,000 journals they were receiving (23). The 300 most cited authors for the period 1961 to 1976 in the *Science Citation Index* were found in eighty-six journals, of which five accounted for more than one-third and ten for about one-half (168). In a bibliography on schistosomiasis covering a 110-year period and containing 10,286 citations, about 50 percent were contained in about fifty journals out of a total of 1,738 cited (457:20). The evidence is so persuasive that Bradford's law has been called "intuitively obvious" (163).

The distribution described by Bradford has been recognized in other kinds of phenomena. When generalized, they suggest that, in any distribution of a large number of occurrences, a disproportionately large number of the occurrences can be attributed to a small number of agents. One of the best-known examples is the law of Vilfredo Pareto, who analyzed the distribution of incomes in the late nineteenth century with similar results. His results were much like those produced by a study of income and expenditures in Great Britain in 1970. The study showed that roughly 80 percent of the income was accounted for by 20 percent of the population (156). A similar distribution was found by Alfred Lotka, who studied the productivity of scientific authors in 1926. He stated that "... the number (of authors) making n contributions is about $1/n^2$ of those making one ..." (281:223). In other words in a population in which 500 authors each published one article, there would be 125 publishing two, 55 publishing three, and so on. Lotka's data were drawn from chemistry and physics, but the phenomenon has been confirmed in many other studies.

Zipf applied the formula to the way authors use words as an expression of what he called the "principle of least effort," which he regarded as a natural law which governs all human activity. He based his conclusions on an analysis which had been made of the frequency of the 26,530 words used by James Joyce in composing *Ulysses.* He found that the rank order of the word in the frequency list multiplied by the frequency remained relatively static along the entire distribution. The tenth word in the list, for example, was used 2,653 times and the word which ranked 1,000 was used twenty-six times (482). This principle has also been used by investigators to try to establish the authorship of anonymous works by comparing word frequencies in writings known to be by the authors and in those attributed to them. Among the examples was the attempt to determine which of the disputed papers in the *Federalist Papers* were written by Madison and which by Hamilton, and whether Thomas à Kempis is truly the author of *The Imitation of Christ.* In fact, these kinds of distributions seem so pervasive that Price was able to generalize them into what he called a "cumulative advantage distribution" (369) which reflects the age-old observation that the rich always seem to get richer.

Questions have been raised about the validity and utility of the studies based on citations to or use of the literature, particularly in examining another phenomenon called obsolescence. Studies of current use, as J. D. Bernal pointed out, do not necessarily represent optimum use, but merely reflect the working habits of current users, and their extent of knowledge about and ability to use the system (34). Most use studies essentially represent a compilation of judgments of relevance which have validity only in terms of the validity of the relevance judgments. Research on the correlation of relevance judgments between different but compatible groups has not been particularly reassuring. Lancaster sums up his observations on studies of use of the literature as follows:

> Perhaps the single most important finding is that accessibility (physical, intellectual and psychological) seems to exceed "perceived value" as a factor de-

Table 4 Percentage of Literature Cited by Age in Years

Subject	0–10	11–20	21–30	31–40	41–50	Over 50
Physics	64	23	8	2	2	1
Chemistry	64	24	6	2	1	3
Physiology	62	18	11	2	3	4
Zoology	26	18	12	7	7	30

Adapted from Vickery (449), 1938–1953 data.

termining which source will be chosen when the need for information arises. (271:380)

The statement underlines Zipf's principle of least effort and provides reinforcement for the idea that propinquity may be one of the most important factors in activities so diverse as selecting items from the literature and choosing a mate. Bradford's "law of scattering," however, does carry one important message, that there are diminishing returns in extending one's searches into the literature beyond a certain point.

Obsolescence and Redundancy

It has been frequently observed that the more recent literature is more heavily cited than the older. This phenomenon has been called obsolescence and defined as "the decline over time in validity and usefulness of information" (276). Along with scattering, which relates to the concentration of use in a few journal titles, obsolescence, the decrease in use of the literature as it grows older, tends to lead to a more compact literature with which the investigator must be concerned. It has been reported that the average citation rate for a scientific paper tends to reach a peak during the third year after publication and to fall off by 50 percent every 3 to 5 years thereafter (295), and that, on the average, half of the citations in any year in the *Science Citation Index* are to the literature of the previous 5 years (371). More precisely, over 65 percent of the citations in the 1981 *Science Citation Index* were to the literature of the last 10 years, and less than 6 percent were to literature older than 30 years (396:29). These rates vary to some extent from discipline to discipline but generally conform to a pattern, with the exception perhaps of the taxonomic sciences, as can be seen in Table 4.

The term "half-life" has been borrowed from nuclear physics to equate the obsolescence of literature with the time required for the disintegration of one-half of a sample of a radioactive substance. It has been defined as the period during which one-half of the currently cited literature was published, that is, a literature with a half-life of 10 years would have 50 percent of its current citations to the literature of the last 10 years. Using this concept, it was calculated in

1960 that the half-life of the literature of metallurgical engineering was 3.9 years, while it was 11.8 years for geology and 7.2 years for physiology (64).

A number of reservations have been expressed about the concepts of obsolescence and half-life of the literature. They are based not on values intrinsic to the literature but on the analysis of citations, that is, individual literature use. Readers tend to refer to the recent literature because as the volume of literature grows there is more recent literature available and because many citations may occur as a result of current awareness rather than retrospective searching. Citation studies may show changes which are due as much to changes in editorial policy on publishing references as they are to obsolescence. It has been suggested that this becomes apparent when one compares the number and kind of citations found in dissertations with those found in publications derived from dissertations (194). Garfield has listed three types of authors who are not cited: (i) the unintelligible, the irrelevant, and the mediocre who don't deserve to be cited; (ii) those who deserve citation but either have not yet been discovered or are forgotten; and (iii) those whose work is so well known that no one thinks it is necessary to cite them. Examples of the last are papers on well-known methods attributed to an individual. He calls these "implicit citations" (169). A similar phenomenon he calls the "obliteration phenomenon" where "important scientific discoveries are quickly incorporated into the common wisdom of the field" (160). He gives as an example an analysis of biochemistry papers published in the 1950s and cited 500 or more times during the period 1961 to 1975. At the lower end of the scale was the Watson-Crick seminal paper on DNA with 552 citations (458). At the top of the heap was a paper by Lowry and others on a method of protein measurement with 50,016 citations (282). This gives credence to the assertion that "methods papers" tend to be cited more frequently than other papers (161).

It has been pointed out that "use of the literature associated with time" is more descriptive than the term obsolescence, which has connotations which are not valid in this context. Obsolescence or decline in use, as Line and Sandison point out in their review which lists over 200 use studies between 1927 and 1974, may be a result of various factors:

- The information is valid but has been incorporated into later work.
- The information is valid but has been superseded by later work.
- The information is valid but in a field of declining interest.
- The information is no longer valid.

They also point out that obsolescence is a reversible process as when:

- The information is considered invalid but later becomes recognized as valid.
- The information is valid but the necessary theory or technology is not yet available in order to exploit it.
- The information is valid but of a low but perhaps increasing rate of interest.

Interests, they add, are sometimes a result of fashions and social pressures. They cite as an example a study of a bibliography on alternatives to gasoline as a motor fuel. The number of references showed great fluctuations for each decade from a ratio of 1 in 1900 to 24 in the 1920s and back down to 2 in the 1960s (276).

There must also be variations and gradations of the obsolescence of knowledge, from the burned-out ideas which have truly been assigned to the "ash heap of history" to those which still have a glimmer of fire in them. Some ideas perhaps may be regarded as monumentally obsolete. An example is the concept of phlogiston, which in the seventeenth century was regarded as an imponderable substance involved in the process of combustion and was not overthrown until late in the eighteenth century by Lavoisier and others. The verb "to overthrow" implies a period of intellectual contention in which an idea may hover at the brink of obsolescence but not yet be pushed over. The history of science is full of such ideas. There is, for example, *Naturphilosophie,* which Medawar defines as "a philosophical indoor pastime which does not seem even by accident (although there is a great deal of it) to have contributed anything to the storehouse of human thought" (300:72). There are also broad concepts such as natural balance or homeostasis which go back to the ancient Greeks and were incorporated into the work of Claude Bernard and Walter Cannon. Obsolescence is not always associated with antiquity. The endocrinologist Bogdanove could say in 1962, "the young field of hypothalmic-pituitary physiology is already littered with dead and dying hypotheses" (45). Nevertheless, it must be disquieting to a student in the life sciences to hear a lecturer say "the half-life of what you are learning is 7 years," particularly if the lecture concludes with a quotation of an aphorism from Hippocrates.

For these reasons the concept of obsolescence may not be very meaningful to the investigator. The concept of redundancy, repetition in the literature of ideas which appear in print elsewhere, may be more significant, but it is even more difficult to identify and to measure. There are, nevertheless, frequent references to it in the literature. Bernal refers to "the abuses of unnecessary, inflated and multiple publication" (34), and Waksman says, "There is also frequent replicate publishing, the same data being published repeatedly in different primary journals" (454), a process which a recent author refers to as "double dipping" (179). There even have been some attempts to quantify it. One author sampled one field in mathematics and determined that 21 percent of the literature represented duplication. He qualifies this by saying that most of it was probably due to "honest sciolism rather than plagiarism" (298). Flagrant plagiarism, in which someone copies another author's work practically word for word and submits it as original, occurs rarely. It is somewhat less detectable when it occurs with only a portion of an original work, or when the original author's ideas are rephrased without appropriate recognition.

There are also forms of inadvertent plagiarism, using someone else's ideas without being aware of their source, which Merton calls "cryptomnesia" (306). This is somewhat different from May's "sciolism," a result of superficial knowledge, cited above. It does not include those cases in which another author's ideas are used without being acknowledged, which is a different and more reprehensible breach of ethics. There are some kinds of redundancy which are unavoidable and even necessary, such as when many ideas or contributions are brought together in different syntheses as in several textbooks on the same subject. In this sense much of the information presented in this book may be regarded as redundant. Additionally, experts tell us that a certain amount of redundancy is desirable in communications systems to ensure optimum transmission of the information.

There is one kind of redundancy which has been measured with some degree of accuracy: that which occurs in secondary sources such as abstracts and indexes which cover related bodies of the literature. One of many examples of this kind of redundancy is a study of the overlap of journals covered by *Biological Abstracts, Chemical Abstracts,* and the *Engineering Index.* Of a total of 14,592 journals covered in the three services in 1970, only 140 were covered by all three, and 10,511 were covered only by one. The overlap was greatest between the *Engineering Index* and *Chemical Abstracts* at 43 percent; it was 32.4 percent between *Chemical Abstracts* and *Biological Abstracts,* but only 1.4 percent between the *Engineering Index* and *Biological Abstracts* (472). Since all these services cover some of the journals on their lists selectively by extracting only those articles which are within their scope, an analysis was also made of the 3,112 journals covered by both *Chemical Abstracts* and *Biological Abstracts.* No overlap of coverage occurred in 1,331 of these journals, since each service extracted different articles. There was complete overlap in 572 journals (23,662 articles) and partial overlap in 1,209 journals (25,194 articles). Of a total of 554,205 articles covered by these services, only 48,856 were covered by both (473).

Writing, Publication Cycle

Another characteristic of the literature on which we have some data is the amount of time it takes from the initiation of a research project to the time it is made available in printed or other formats. It also relates to the factor of obsolescence, because obviously the longer it takes to get information into print, the greater the likelihood it may be obsolete when it reaches its destination. The process is more important in science than it is in the humanities, such as in historical writings where it is possible to reach different interpretations from the same data. Interpretations from the same data are expected to be the same in the sciences, so it is more likely that duplication of effort can occur if information is not disseminated in a timely fashion (299:36). This is why statements such as "The average elapsed time from the initiation of work to the publication of a journal article is about two to two and one-half years, depending on the field" (246:5) can be disquieting. Of course, many things can happen during this interval. The work may be informally discussed in various groups or with various individuals. Oral reports may be made at local or special group meetings or at state, regional, national, or international conferences. It may also

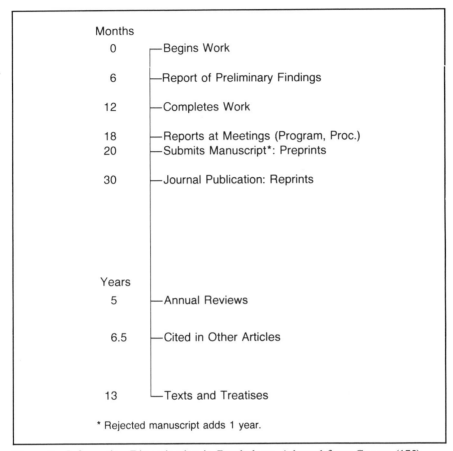

Figure 1   Information Dissemination in Psychology. Adapted from Garvey (175).

appear in the form of an abstract prepared for one of these meetings or be distributed in the form of preprints.

An extensive survey of this information dissemination cycle was made for the American Psychological Association about 1970. Since the average elapsed time from submission of a manuscript to journal publication has been estimated to be 12.1 months for both psychology and the life sciences (246:68), the information may be considered relevant for the other parts of the cycle as well. The investigation showed that the steps between the inception of the work and journal publication take on the average 30 months (see Fig. 1) and that it may take as long as 13 years before it is incorporated into a textbook. About three-fourths of those who submit reports to the American Psychological Association's annual meeting publish in the following 3 to 4 years. Articles in psychology tend to get published in abstract services about 8 months after they appear in print, but it is apparently 2 to 3 years before they are covered in reviews. It is about 4 years before they are cited with any frequency in reviews (179).

It is for reasons such as these, says Waksman, that meetings have tended to replace journals as media of communication in the sciences. There were eighty-nine Gordon Conferences on special topics held in the summer of 1979 alone, he reports, and over fifty workshops by the National Institutes of Health during the same period (454:1011). It has also been estimated that 50 percent of all scientific authors make one or more prepublication reports, with two-thirds making oral reports and one-half making written reports (246:5). Orr followed the papers presented orally at the meetings of the American Heart Association in 1958. He found that only 41 percent had been published before the end of the year, despite the fact that one-third had been submitted to a journal before the meeting (342). The lag time between submission of a manuscript to a journal and publication seems to vary considerably from field to field. The range in one study was from 8 months for the physical sciences to 20.5 months for mathematics, with life sciences, as I indicated before, near the middle of the range at 12.1 months (246:68). This is a matter of concern particularly to investigators who depend on publication to establish their priority. Many journals have recognized this concern by indicating the submission date on the published article.

These characteristics of the scientific literature are all relevant to libraries in deciding which journals to add to or retain in their collections and to individual investigators in choosing the part with which they need to be concerned. These characteristics cannot, however, be the final determinant because they do not always have predictive value or decide how any individual item of information will be used no matter what its origin, language, age, or source, and because choices must eventually be based on qualitative judgments.

Chapter 6

Reading the Literature: Selection, Evaluation, and Access

Selection

Scientists have complained for centuries about their inability to "keep up with the literature." Historically, they have tried to solve this problem, first by narrowing their fields of interest and second by being more selective in what they read, by trying to pay attention only to those sources which are most productive of new and reliable information. "It is a truism," says Waksman, "that nobody can get through all the published pages in a field like immunology in the time available." He points out that there are thirty-five primary journals and twenty-seven review series which are current in that field, and that twenty-three of the primary journals had their origins in the years 1970–1979. The *Journal of Immunology* alone now publishes over 5,000 pages a year. Moreover, he adds, ". . . the proportion of really innovative work appears to be diminishing; as the limits of knowledge are extended, an even higher proportion of published work is concerned with details rather than principles" (454:1011). In 1967, the advice was already being offered that: "As a result of the recent expansion of scientific literature, more time and effort are being devoted to the selection of what is to be read than to the actual reading" (295). The American medical educator George E. Miller made the point in his comment on "speed reading" courses. The problem of coping with the literature, he said, was not one of learning how to read faster but how to select what was worthwhile reading slowly. Many years before this, the American humorist Finley Peter Dunne's (1867–1936) character Mr. Dooley responded to exhortations to read more by reminding us that "readin' ain't thinkin'."

The methods one uses to select literature will vary considerably with individual needs and purposes. There are two primary ways we use the literature. One is to discover what is new that is relevant to our interests. This mode is sometimes called browsing, a descriptive word of uncertain origin which refers to methods of scanning literature. It is also referred to as maintaining current awareness. The other purpose is to solve a particular problem or to answer a specific need for information. The two modes require quite different techniques. In fact, it has been suggested that the literature itself should be divided into two parts: a current awareness literature which serves "to inform, interpret, criticize and stimulate," and an archival literature where new experiments and observations which form the basis for the advance of knowledge are recorded (457:22). Relman divides the literature into three types: (i) articles presenting new research, laboratory, or clinical data which add new information, (ii) teaching or didactic articles which can report previous research (review articles are in this category), and (iii) analytical, speculative, and editorial articles whose intent is to express personal opinions or to clarify ideas (379:67).

In this chapter I am concerned primarily with selection and reading for current awareness. The process of identifying literature to solve a specific information problem is dealt with in Chapter 10 "Searching the Literature." There the first step is to clearly articulate the problem you wish to solve and to develop a specific strategy for solving it.

Periodic scanning of the literature and routinely following certain journals are not the only methods of maintaining current awareness. Various agencies supply "selective dissemination of information" (SDI) services, most of which are computer generated. Some vendors permit subscribers to follow a specific subject on a weekly, biweekly, or monthly basis. The service is based on a search profile developed in consultation with the user and is processed periodically as the database is updated. These services are offered by most of the database vendors covered in Chapter 10, as well as by database producers such as BioSciences Information Service, Chemical Abstracts Service, the National Library of Medicine, and the Institute for Scientific Information.

Another form of current awareness which serves as an alternative to finding and scanning large numbers of journals is provided by *Current Contents,* which is published by the Institute for Scientific Information. There are now editions in seven disciplines, including one in the life sciences. *Current Contents Life Sciences* covers the most important basic science and clinical journals as well as 500 multi-authored books each year. Each weekly issue consists of reproductions of the contents pages of the journals covered. Foreign language titles are translated into English, and an author list with addresses is added to facilitate requests for reprints. There is also a list of all the significant words in the article titles, which serves as a subject index. Evidence of the heavy use of *Current Contents* by the scientific community was provided by an author whose article title was transcribed with minor errors in one issue. He tallied his reprint requests for this article into three categories: (i) those which contained the correct title and were therefore derived from sources other than *Current Contents,* (ii) those with the incorrect title as it appeared in *Current Contents,* and (iii) those which could not be classified in either category. He found that 85 percent of the reprint requests fell into the second category (341).

It is often recommended that all scientists should have a journal reading plan most appropriate to their needs, interests, and available time. The number of journals you elect to pursue on a regular basis will depend not only on these factors, but on the concentration or dispersion of articles in your subject. A "realistic journal reading plan" for medical practitioners has been suggested to consist of two weekly peer-reviewed journals, two monthly refereed journals, and two standard "newsletters" (393). Most strategies recommend selecting a relatively small nucleus of basic, essential, or core journals to follow on a regular basis. One "conservative" estimate is that a research worker or physician who is interested in the laboratory and clinical aspects of heart disease needs to follow some seventeen key journals. The author admits, however, that a more reasonable limit is probably about ten journals a month (39). The editor of the *New England Journal of Medicine*, A. S. Relman, informs us: "To stay well informed, the average practitioner only needs to read a few, well-chosen periodicals, one or two general medical journals (weeklies) plus two or three specialty journals (usually monthlies)" (379:77).

In some fields the problem of selecting the appropriate journals is not an easy one. The National Academy of Sciences in its survey of the life sciences reported: "In almost every scientific subfield there is a hierarchy of journals that reflects the relative quality of published papers. Although it does not exist overtly, this hierarchy is known to all sophisticated scientists within the field." (318:43) Selection, therefore, is based on the personal and collective opinion of workers in the field. One method of eliciting or discovering this consensus is the frequency with which papers in particular journals are cited. Each year the Institute for Scientific Information publishes *Journal Citation Reports*, which provides detailed information of this kind, from which core lists of journals in many fields can be determined. I will discuss this publication at greater length in Chapter 9, "Citation Indexing and Analysis."

Some individuals who wish to be more systematic maintain checklists on which they keep a record of the journal issues they have scanned. There are preprinted forms which provide spaces where weekly or monthly issues can be checked, but it can be accomplished by using pages in a notebook or cards for each title on which volume and issue are entered as the journals are scanned. Many research libraries display their daily or weekly journal receipts in a special place for a short period before they are filed with the other issues. By visiting this area on a regular basis it is easy to see all the journals on a core list as they are received.

Reading

There is much advice in the literature about how to scan journals and journal articles and how to read them critically. Gehlbach in his book *Interpreting the Medical Literature* cautions us to avoid compulsive behavior in reading the literature despite the guilt feelings aroused by seeing piles of unread journals. "We need," he says, "to develop tactics for sampling journals and consuming only those articles that are most nutritious" (178:10).

As Mortimer Adler suggests in his *How to Read a Book* (9), knowing the structure of a literary work is important in reading it effectively. Fortunately, most journal articles in the life sciences have similar structures.

1. They usually begin with an *abstract* or summary which is a concise statement of why the study was performed, how it was carried out, and what the results were. It provides a quick way to determine whether or not to read further. Waksman says ". . . it is clear that the real articles are now the abstracts; the rest of each article is a technical report, available if needed, but rarely read" (454:1011).
2. The *introduction* provides some of the historical background and the reasons for the study.
3. The *methodology* section is important in determining the study's validity, although, says Gehlbach, it is most frequently skipped in reading. Once it is determined that an article is relevant, he suggests, the methods section should be read first. "Here," he says "is the substance of the research." Here the sampling methods, the study design, the data collection techniques, and the evaluation procedures are described. If the methods are not sound, the interpretation and conclusions are not worth reading (178:12).
4. The *results* section presents the findings of the study in narrative form, tables, graphs, or figures. Analysis and interpretation may also appear here.
5. The *discussion* reviews the study and presents conclusions which can be derived and sometimes suggestions for further work. It may also, Gehlbach adds, include "disclaimers, equivocations, apologies, chest thumping, speculation, instructions, fantasy" (178:7).
6. Finally, the *references* or bibliography indicates the extent to which the author has consulted the literature. It may be the most useful part of the article since it can lead to additional relevant material.

Developing a strategy for rejecting articles for reading is important, says Sackett. "It is only through such early rejection of *most* articles that busy clinicians can focus on the *few* that are both valid and applicable in their own practice" (390:124). The procedure he suggests is to (i) look at the title to determine its relevance, (ii) examine the authors in terms of what you know of their reputation, (iii) read the summary to determine whether the conclusions are valid and important, and then (iv) consider how it relates to your current needs. It is only if it passes all these tests, he advises, that you should proceed to read the section on methodology.

Systematic screening of the literature should not preclude browsing, which can be described as a form of wandering at random through the literature for both pleasure and profit. Of the three methods of using the periodical literature (current awareness, problem solving, browsing), says Relman, "Browsing is the most fun. It is the least efficient of the three methods and yet invaluable as a catalyst for new ideas and fresh perspectives" (379:76). Bentley Glass asked fifty investigators how they located five of the most important papers they had cited in one of their recent publications. Nearly half, they said, were discovered through casual browsing or from a journal they regularly scanned (51:260).

Evaluation

We are all somewhat conditioned by our reverence for holy writ to believe that anything in print is gospel. Ziman points out in his *Public Knowledge* that even accepted scientific findings and theories rest in part on faith and that they sometimes require as much proselytizing as ideas in religion (476). Many errors of scientific bias are not easily detectable except through replication of the experiments involved. "Our current literature," says Chargaff, "is brimming with facts, but many, I am afraid, are no longer available" (74). There is, however, not much incentive to replicate scientific experiments because little or no prestige is associated with reproducing someone else's results (57:60). Incorrect results instead are usually allowed "to fade into obscurity" (299:45). Some research cannot be replicated, says Collins, because the "algorithm," the procedures, are not fully documented, because of "competitive secretiveness, incompleteness of the article and so on" (82). Also, some studies, particularly those which involve large population samples, are expensive or troublesome to replicate.

Responsibility for the quality and validity of an article rests to some extent on the editor of the journal and those enlisted to review it. The responsibility ultimately, of course, rests with the authors, and journals sometimes include disclaimers to make this clear. The reader, however, shares in the responsibility by reading critically and by not accepting everything read. Typographical errors, for example, do occur. It has been suggested that spinach got its reputation as a dietary supplement because of a misplaced decimal point in which the iron value was given as ten times higher than it was (203). An interesting case was reported in the French medical literature under the title "Mortal Consequences of a Typographical Error" in which the reader was actually held culpable. A German physician had read in a journal that cases of *pruritis ani* had been successfully treated with a 1 percent solution of percaine. When he used this solution in treatment, his patient died. He was not aware that the editor in a subsequent issue of the same journal had published a correction stating that an error of one magnitude had taken place and that the solution should have been one per thousand. A German court, nevertheless held the physician guilty of negligence on the grounds that, considering the novelty of the treatment, he should have followed the literature more closely and sought advice from competent authorities (87).

It is in analysis of the research design and evaluation of the statistical methods which have been applied that critical reading comes into play. No matter how well executed the research study, says Altman, if the statistical part is below standard, the research loses much of its value and may even be considered unethical (12). Several authors have indicated that some research studies are deficient in these aspects. "Approximately half the articles published in medical journals that use statistical methods," says Glantz, "use them incorrectly." He cites other studies of the research literature which confirm his analysis (184). A study in 1966 of ten of the most frequently read American medical journals reviewed 149 research studies and reported that only 28 percent could be regarded as acceptable. The deficiencies included:

- absence of a control group, e.g., administering a treatment to one group without comparison to a similar group which did not receive the treatment
- use of statistical techniques which were not appropriate to the data or the way in which the data were collected
- conclusions drawn which were improper for the statistical tests used
- use of study designs inappropriate to the problem
- failure to use statistical tests when required.

The most frequently occurring error was conclusions drawn about a specific population without reference to any statistical test (394). A study of the *British Medical Journal* 10 years later found statistical errors of one kind or another in over half of the research reports studied (189).

Criticisms have been applied particularly to controlled clinical trials in which new drugs or treatments are evaluated in accordance with well-defined protocols. One study analyzed 5,737 articles published between 1972 and 1979 relating to the products of a single pharmaceutical house. Only 61 percent of the articles were found to include such essentials as the number of patients treated and the number of adverse reactions, and only 19 percent were considered to have included all the required data. No particular medical specialty, said the authors, seemed to have a better record than any other (448). Feinstein reviewed the articles in five English-language general medical journals for a 6-month period in 1974 and reported that only one-third of the scientific papers contained statistical procedures. He concluded that "... the most important issues in biostatistics are not expressed with statistical procedures. The issues are inherently scientific ... and relate to the architectural design of the research" (135).

Four design features have been identified as essential to clinical trials but are relevant to other research studies as well: (i) controls, (ii) randomization, (iii) objective measures, and (iv) statistical analysis. Mosteller reports a study of clinical trials in which almost half had none of the above elements and in which only 21 percent reported use of controls (313).

To read the research literature critically, therefore, one must have some awareness of how to evaluate research design, statistical procedure, and the validity of the data. A large literature on the design of experiments and on statistical methodology is readily available in most research libraries (17, 134, 216, 224, 315, 382, 386, 442). The problems identified include omission of information and failure to recognize sources of bias. Among the omissions which Mosteller finds are (i) details about the methods of randomization, which are important in evaluating research results, (ii) failure to report the statistical devices used in the analysis, and (iii) failure to report the descriptive statistics on which tests of significance are based (314).

Sackett has identified as many as thirty-five different forms of bias in sampling and measurement in research studies. He was able to catalog them under six headings:

1. Biases of rhetoric where persuasive language is used instead of sound argumentation and where only favorable examples are chosen to support a thesis
2. Biases in specifying and selecting the sample (the richest source of bias, with twenty-two listed), including inappropriate sample size and "membership bias," i.e., generalizing to a larger population from the membership of a single group
3. Biases of methodology such as omitting withdrawals from the results and failure to monitor compliance in subjects
4. Biases influencing measurement such as "apprehension bias" in which the subject's reaction may influence the results,
5. Biases in the analysis such as excluding data which do not fit desired statistical patterns, which he calls the "tidying up bias"
6. Biases in interpretation such as equating correlations with causes (389).

Bias is in fact often difficult to identify and to avoid. "There is no such thing as unprejudiced observation," Medawar said in answering the question: "Is the scientific paper fraudulent?" It embodies, he said ". . . a totally mistaken conception, even a travesty of the nature of scientific thought." Research is a process where sufficient data are collected from which generalizations can be developed inductively. It is almost always, he said, based on deductive reasoning in which, starting from a position of bias, one sets out to prove or disprove a hypothesis by collecting relevant data (301). In another one of his essays he describes research as a delicate balance between the imagined and the observed.

> All advances of scientific understanding, at every level, begin with a speculative adventure, and imaginative preconception of *what might be true*—a preconception which always, and necessarily, goes a little way (sometimes a long way) beyond anything which we have logical or factual authority to believe in. The conjecture is then exposed to criticism to find out whether that imagined world is anything like the real one. Scientific reasoning is therefore at all levels an interaction between two episodes of thought—a dialogue between two voices, the one imaginative and the other critical; a dialogue, if you like, between the possible and the actual, between proposal and disposal, conjecture and criticism, between what might be true and what is in fact the case. (302:532)

Evidence can be found in the history of science in which the balance is sometimes weighed in the direction of imagination rather than observation.

Frauds, Hoaxes, and Plagiarism

Incidence of statistical error is not the only reason to read the scientific literature carefully. In addition to honest error and inadequate experimental design, cases of conscious misrepresentation have occasionally been reported. There are many tales of frauds, hoaxes, and plagiarisms. Not even the famous have been exempted. In their catalog of "known or suspected cases of scientific fraud," the authors of *Betrayers of the Truth* include Ptolemy, the Greek astronomer whose work formed the basis of astronomical thought for 1,500 years, Galileo, Isaac Newton, and Mendel (57:225). The celebrated German zoologist Ernest Haeckel (1834–1919), who promoted the theory that the embryo retraces its evolutionary history in utero, altered the illustrations in one of his papers by labeling three copies of the same plate as embryos of human, dog, and rabbit to prove their similarity. When accused, he defended himself by saying ". . . hundreds of the best observers and biologists lie under the same charges" (203). Recently, Gerald Geison has implied that even Pasteur can be added to the list. Using information from Pasteur's laboratory notebooks which have only recently become available to investigators, he discovered that Pasteur may have been guilty of "creative reporting," since his laboratory notes differed in several significant ways from his published reports on the anthrax and the rabies vaccines (274).

There are no data on the frequency with which this phenomenon occurs. Reviewers of *Betrayers of the Truth* have been critical of the level of research on which it is based, calling it a journalistic account with a rather loose, inclusive definition of fraud. One reviewer comments:

> Since there are about one million scientists alive today and approximately one fraud is revealed each year, the incidence is once in 10,000. Few human endeavors have so low a noise level. (479)

Evidence, however, emerges from time to time that fraud occurs even at highly respected institutions. The level of outrage that follows the confirmation of wrongdoing demonstrates how strongly it is against the mores of the entire scientific community. An attempt was made to quantify the extent to which cheating or insertion of intentional bias occurs in scientific writing, through a questionnaire directed to the readers of a large-circulation general scientific journal. They were asked to respond on their personal knowledge of such cases. The results can hardly be considered scientific since only 204 replies were received, including one signed by "a laboratory rat" and another by "an inspector of custard pie stability." Of the respondents 92 percent claimed to have some direct or indirect experience of intentional bias in reporting research results; 66 percent reported experience of more than a single case (391). At a meeting of biology editors in Boston in 1981, the audience was asked if any of them knew of unpublicized cases of fraud, and over one-third raised their hands (13).

Hoaxes sometimes start off as innocent jokes, but some leave lasting impressions in the literature. A classical example is Henry Mencken's bathtub story, which was first published in the *New York Evening Mail* in 1917. He stated authoritatively that the first bathtub in America was installed in Cincinnati in 1842 and that there was not one in the White House until it was added in 1851 at the request of President Millard Fillmore (234). It was widely quoted and reprinted as fact, and even the noted American bacteriologist Hans Zinsser reported in 1938 that "the first bath tub did not reach America—we believe—until about 1840" (480:285).

The celebrated medical clinician William Osler was addicted to pranks of this kind. He invented Egerton Y. Davis as a pen name to cover the jokes and hoaxes he planted in the literature. A list of eighteen items assigned to E.Y.D. as author has been compiled (431). His earliest effort in this vein occurred when, while still a medical student, he published a case of "penis captivus" or *de cohesione in coitu,* a condition which is now reported to have only hearsay evidence and "to have vanished perhaps completely in this century." Osler's case was completely fabricated. It is nevertheless cited in Kisch's *Sexual Life of Women* as an actual case reported by a "medical man called Davis, not otherwise identified" (429). Osler's humor often had a Rabelaisian bent. Another of his hoaxes is a reported case of "Peyronie's disease or *strabismus du penis*" in which ". . . when erect it curved to one side in such a way as to form a semicircle, hopeless and useless for any practical purpose." This time he signed his contribution with the name of a prominent Philadelphia urologist, who responded by confirming the observation and signing it EYD Jr. (317).

Some cases in the history of science seem to lie in a middle ground between hoax and fraud. In the well-known case of the Piltdown man, a carefully modified ape's jaw was buried in an English gravel pit in 1912 along with a human skull to simulate the remains of an early ancestor. It engaged the attention of a number of eminent anthropologists and zoologists until it was finally disclosed as a hoax in 1953 (191:201). The case of the "midwife toad" still excites controversy (255). Paul Kammerer was an Austrian zoologist who claimed in the 1920s to have induced the transmission of acquired characteristics and thus provoked the orthodox Darwinians of the day. He claimed to have demonstrated this by means of the midwife toad which breeds on dry land and therefore never develops "nuptial pads" which are characteristic of a related species that breeds in water. By breeding several generations of the dry land toad in water, he claimed to have induced the transmission of the ability to develop nuptial pads from one generation to another. A visitor to his laboratory discovered that one of the specimens had been injected with india ink at the point where the nuptial pads were said to exist, and Kammerer was accused of fraud. It has never apparently been proved that Kammerer was the actual culprit, but the story has a tragic end since he committed suicide.

In recent years a number of cases of fraud and misrepresentation of research results have been reported from several well-known laboratories. They include:

- a professor at Stanford who was found in 1973 to have published a number of articles which included citations to papers which had never been published (418)

- an investigator at the Sloan-Kettering Institute who admitted in 1975 that he had used a black felt-tipped pen to simulate a successful skin graft on a white mouse (à la Kammerer) (57:153)
- a junior investigator at Yale who committed both plagiarism and fraud in 1980 by falsifying work in a dozen papers he co-authored with the head of his department (58)
- a young investigator at Harvard who had impressed his superiors that he had a distinguished research career ahead until it was discovered that he had committed a large number of scientific frauds back to his undergraduate days (251).

The unsuspecting reader has little protection against these counterfeits except sharp perception and reports of exposures. Retractions, however, seldom occur, and when they do they are sometimes couched in ambiguous terms (136). I once suggested somewhat facetiously that a journal be initiated under the title *Journal of Fraudulent and Plagiaristic Research* in which reports of this kind could be exposed. The objection was raised that there was little difference between fraudulent and inaccurate research and that in many cases the determination might be difficult, no matter what the motivation. Fortunately, more and more attention is being paid to ethical concerns in research and publishing. The research enterprise, says Relman, is based on trust and editors "have no choice but to assume that the authors have honestly reported what they did and what they observed." It is in fact, he says, impossible for collaborators, referees and readers to verify every datum that is reported (380). Suggestions have been made such as providing a uniform format for retractions and assigning more responsibility to referees of journal articles by dropping their cloak of anonymity. Recently, more retractions have been published (102), and efforts are being made by official scientific bodies to establish guidelines on research fraud and misconduct and to develop recommendations to prevent them from occurring (14).

Plagiarism raises ethical issues other than those concerned with the falsification of data. To the reader the validity of the data may be the primary concern. However, an author guilty of plagiarism may be regarded as suspect in other ways as well. Real and alleged examples of plagiarism have existed since the early history of science. The issue is clouded by questions of when plagiarism can be said to exist. It is more discernible in general philosophy and humanistic studies, says Ben-David, because:

> ... style is so much a part of the message that plagiarism requires actual copying. Actual copying betrays itself. ... Scientific results, because they are specific and are independent of writing style, are easy to steal. Moreover, because of the frequency of genuinely independent simultaneous discovery, it is difficult to detect plagiarism or, in the case of genuine multiple discovery, to assign property rights. (29:249)

Plagiarism can even occur, it has been observed, without the knowledge of the author. Merton uses the term *cryptomnesia* (which has been attributed to

Freud) (306) for this kind of "unconscious plagiarism." He cites as an example the mathematician William Rowan Hamilton, who developed a lifelong preoccupation with unknowingly reporting someone else's ideas as his own. "As to myself," he said, "I am sure that I must have often reproduced things that I have read long before without being able to identify them as belonging to other persons" (306:404).

The issue is complicated, as Ben-David says, by the well-known phenomenon of simultaneous or duplicate discoveries in which two or more individuals arrive at a solution of a problem at the same time (29). There are many such examples in the history of invention. In the life sciences the independent development of the theory of evolution by Wallace and Darwin is well known. Ogburn lists 148 alleged cases of this kind which have occurred in all fields of science, of which very few can be demonstrated to be the results of plagiarism (337). Plagiarism occurs at many levels from outright use of a report in the author's own words to the simple failure to cite a reference to work from which an idea or data have been derived.

There is a long history of accusations of plagiarism among scientists, particularly in earlier periods when less restraint was exercised in these exchanges. John Flamstead, the Royal Astronomer in seventeenth-century England, for instance, referred to Edmund Halley, who gave his name to the comet, as a "lazy and malicious thief" (306:404). One of the most flagrant cases of plagiarism in the history of science is perhaps that of Everard Home, to whom John Hunter's papers and notebooks went after Hunter's death. Home read 143 papers at the Royal Society, more than any member in its history, and was referred to as "another Hunter." "The comparison was apt," says one of Hunter's biographers, "since most of the original materials had been lifted from John's papers." Home did acknowledge Hunter in some of his early papers, but later neglected to do so entirely. One paper he published as his own, unknown to him, had been published by Hunter 27 years earlier in the *Philosophical Transactions* (252:319).

One of the most flagrant modern cases of plagiarism involved a foreign investigator who worked at several well-known research centers in the United States. Over a short period of time around 1979, he published over sixty articles in Japanese and European journals. Many of these, it was discovered, had been copied almost word for word from other journals, grant applications, and dissertations which came to his attention (57:208). On the basis that "imitation is the sincerest of flattery," plagiarism may not be a concern to the reader except for the moral issues it raises. Cases have, however, been reported in which plagiarism was combined with fraudulent data, so that any instance of plagiarism can cast suspicion on the general credibility of the author.

Secrecy

Another element over which the reader has little control but which provides a barrier to obtaining correct and complete information is secrecy. The progress of science is based on cooperative and cumulative effort founded on the free exchange of ideas. The ethical considerations in withholding information nevertheless are by no means clear and unequivocal. In some cases it has been suggested that delaying or withholding publication may be desirable. "It has been estimated," one author says, "for example, that some 90% of new drugs described in newspaper reports over the past 20 to 30 years either never reached the market or are no longer on the market because they are ineffective or toxic or both" (439).

Information is not withheld, however, only because it may be erroneous or because inadequate evidence is available to support it. Many cases are accounted for by the anticipated danger of having one's ideas plagiarized through prior disclosure and thus being deprived of the scientist's primary reward, the recognition for having generated a new idea, fact, or discovery. In commenting on the ambivalence of scientists toward a shared value of free exchange of ideas, Merton says:

> . . . the values and reward-system of science, with their pathogenic emphasis on originality, help account for certain deviant behavior of scientists: secretiveness during the early stages of inquiry, lest they be forestalled, violent conflict over priority, a flow of premature publication to establish later claims to having been first. (305:81)

The history of science, therefore, is full of accounts of disputes over priority as Merton has demonstrated.

> . . . whenever the biography or autobiography of a scientist announces that he had little or no concern with priority of discovery, there is a reasonably good chance that, not many pages later in the book, we shall find him deeply embroiled in one or another battle over priority. (306:385)

Methods of protecting priority without disclosing the complete results of one's research have been practiced for a long time. The *Académie des Sciences* in eighteenth-century Paris permitted a member to present a manuscript to the secretary who initialed and dated it, and retained it in his files until the author was ready to read it. In this way Lavoisier registered his 1772 paper which led to the revolutionary concept of the nature of combustion: "Memoir on elementary fire." This memoir is recorded in the minutes of the *Académie* but apparently was never read (197). Lavoisier's correspondence fully acknowledges that his motives were to protect his priorities, an action which may be a result of the fact that Priestley was hard on his heels (182:61). These methods have continued in modern times albeit in different forms such as leaving out essential details in reporting laboratory reaction procedures. Information, for example, on which solution to use first in a reaction may cause it to fail if the order is reversed. Another method is through the publication of abstracts or letters in such journals as *Science* or *Nature* in which only partial disclosure is made (201). This kind of secrecy can also be circumvented, as Fabregé suggests by reporting on a case in which one worker was refused a request for a sample of a bacteriophage from a laboratory but succeeded in culturing it in his own laboratory by incubating the refusal letter. The first laboratory, Fabregé adds, may now be disinfecting all outgoing mail (132).

An example of the secrecy that can result from intense scientific rivalry is depicted in Wade's *The Nobel Duel* in which he describes the events which led up to the joint award to Guillemin and Schally for discovering the composition of the thyrotropin releasing factor (453). A sociologist who studied the course of events leading to the discovery reported that neither group of scientists (each of the scientists represented the efforts of collective research) yielded precedence to the other. "Instead," he says, "each claimed that the other had made the discovery later than themselves and had received credit by virtue of deliberate ambiguities in their accounts of the investigations" (272:112). The term "deliberate" implies some kind of dishonesty or chicanery, but to a degree reflects the way secrecy can be employed to advance some of the nonscientific goals of the investigator, sometimes for effect and drama as when Schally withheld the announcement of one of his findings because he wanted to make it at a time and place of his own choosing.

These issues have become more visible and more complex with the increasing involvement of industry and government in academic research where "questions of sovereignty and secrecy and of proprietary rights over research have been raised" (326). Questions on who controls the data, the investigators who produce them or the institutions which support their research, relate to the confidentiality of the information, the rights to reap the benefits, and other issues. Fabregé points out that, while there are laws which regulate the dissemination of information in business such as reports to stockholders, there are none in science, where practice is based largely on ethical considerations and tradition (132). In recent legal cases court decisions have upheld the right of researchers to withhold information in some cases but denied it in others.

Withholding information and secrecy in military and governmental research may relate to national security and pose their own requirements. Fears have been expressed that a recently signed federal executive order which prescribes a system for classifying information gives government officials unprecedented authority to intrude whenever they wish to control federally supported research. It has been suggested that "classification, or the worry that it might be imposed, could result in isolation of academic researchers, cut off the free exchange of ideas and exposure to constructive criticism" (384). On the other hand, the Freedom of Information Act, which outlines public rights of access to government records, including those resulting from government-sponsored research, has been challenged in several cases, particularly those in which it is claimed that the confidentiality of information obtained from research subjects should be respected.

Industrially sponsored research has its own claim to secrecy because of industry's desire to recover its investment in the research effort, particularly where exclusive licenses can be granted to any patentable development (100). The recent growth of commercial interest in biotechnology is a good example.

In discussing a new publication in this field, a reviewer said: "It may seem remarkable to some that *Biotechnology Letters* can function at all when it often appears that anyone with a genuinely good idea in this area first telephones their patent attorney and then contacts a banker" (338).

Patterns of behavior in the dissemination of information among scientists, it has been observed, tend to seek the least common denominator. In a situation where a secretive scientist benefits from an open scientist, the open scientist will learn to be more secretive (132). For this reason professional societies have been making efforts to establish guidelines on issues relating to priorities and disclosure of information (326). In the area of industrially sponsored academic research, universities are also beginning to define the obligations and responsibilities under which they are willing to participate, particularly as it relates to restrictions on the dissemination of information (180).

Access to the Literature

Of the three aspects of reading the literature—selection, evaluation, and access—I have discussed the first two. Neither of these has any meaning, however, unless the literature can be acquired. Libraries should be the most important source, even though few of them can afford to subscribe to all the journals required. It is not unusual, nevertheless, to find libraries in the life sciences which contain 2,500 or more journal titles. Those titles not in their possession can usually be obtained from other collections in a few days through various networks. Surveys indicate that the major source of articles read by American scientists, however, is their own subscription copies (69 percent) and that only 14 percent are from libraries. The data for the life sciences varied in that 50 percent were from subscription copies and 26 percent were from libraries. The data for reprints as a source also varied from 2 percent for all sciences to 8 percent for the life sciences (246:166).

The term "reprint" is used for separate copies of an article made by the publisher at the same time it is issued in a journal. They are usually printed at the expense of the author or in return for paying "page charges," a subsidy provided by the author to the journal which publishes the article. The term "offprint" is sometimes also used for this form of publication, but it should not be confused with the "preprints" or manuscripts of articles which are distributed either before or after the acceptance by a journal, but before they are printed. One survey indicated that somewhat over half of the authors distributed preprints, with an average of slightly less than nine for each author (246:6). Reprints are sometimes distributed in larger numbers by authors, either to a list of individuals they wish to make aware of their work or in response to requests. The distribution of reprints can be a measure of the numbers of participants in a subject field. It was reported that the average number of reprints distributed by a single author was highest in the life sciences, with 110, and lowest in mathematics, with 21 (246:75). Some publishers distribute reprints in response to requests, and these may account for about half of all those distributed (246:113).

It is reported that reprints are no longer as widely distributed as they were in the past. Although large numbers of requests may be salutary to the author's ego, they can sometimes impose a large financial burden. One investigator complained that it would have cost him over $5,000 to purchase, address, and mail all the reprints that were requested for a single article (253). The privilege is sometimes abused by those who order reprints without determining that they really need them or by those who are habitual and systematic collectors ("scientific pack rats") (235). As a result of these pressures, and the fact that photocopying is frequently quicker and more convenient, individuals are turning more and more to photocopy machines in libraries as a source of copies.

Copies of articles are also available from other sources such as commercial vendors, some of whom accept online orders through various computer networks and can be paid by means of credit cards. University Microfilms in Ann Arbor instituted such a service in 1983. Agencies such as Chemical Abstracts Service accept orders by telephone, by telex, or through their online service. The "Original Article Text Service" (OATS) of the Institute for Scientific Information supplies order cards which can be used for requesting reprints, many of which are supplied in original form for any of the over 6,800 journals indexed.

The cost and means for providing access to literature are subject to change in this world of rapidly changing technology. Facsimile transmission, in which text is sent over regular telephone wires, after many years is finally becoming a cost-effective method of delivering documents. Many journals are already available electronically online, can be offprinted at a computer terminal, and may very well provide one of our primary methods of delivering documents on demand in the future.

Language Barriers

Another barrier to reading may be imposed by foreign languages. Foreign languages are becoming less and less of a problem as more articles are being published in English. Some countries with other vernaculars, such as German, Swedish, Norwegian, and Japanese, are issuing many of their scientific journals in English. A biochemist made the somewhat exaggerated prediction in 1970 that, if the trend then current continued, virtually all biochemical research would be published in English by 1975 (459). Nevertheless, a significant portion of the world's scientific literature continues to be published in languages other than English, one reason being that some articles may be addressed primarily to the author's compatriots. An analysis in 1971 of 4,800 Japanese scientific journals showed that only 10 percent were published in English, 7.5 percent included a few articles in English, and only 18 percent provided English abstracts (51:40). It has been estimated that about 50 percent of all scientific publications are in languages other than English and that the proportion is tending to increase with the growing prosperity and national awareness among non-English-speaking countries (43).

Attention has repeatedly been called to "Americans' incompetence in foreign languages." The dominance of English in the scientific literature is cited as one of the reasons, as well as the abandonment of language requirements for advanced degrees (367). The situation is somewhat different in England, where a survey showed that over 90 percent of the scientists studied claimed some degree of competence in French and 66 percent in German, although few claimed fluency. Only 10 percent claimed to be able to cope with Russian and fewer than 0.5 percent with Chinese or Japanese. Over 75 percent of the sample, nevertheless, said that they had recently encountered a paper in a foreign language which they could not read because of the language barrier (471). As a result of these deficiencies, access to translations of the foreign literature is frequently required.

Acquiring facility in another language is, as most of us know, a difficult and arduous task. Translation requires skill and facility in two languages. Although scientific translations are under no obligation to be beautiful, faithfulness to the original is highly important. In addition to having an excellent command of one's own language, and fluency in the other language, scientific translators must have knowledge of the subject matter. Good translations are, therefore, difficult to obtain and expensive to purchase.

There were high expectations that computers would help to solve the translation problem when it was recognized that they had extraordinary ability to match words in one language with their cognates in another. Research on machine translation (MT) has been going on for over 30 years, and although occasional announcements are still heard that "machines break down language barriers" (287), it is generally agreed that they have a limited role to play in the process. The structure of language, the subtleties of meaning, and the dependence of words on context have proved to be far greater than early enthusiasts imagined. An old anecdote illustrates the problems. A group of engineers claimed to have built the perfect translation machine. They were asked to feed the biblical phrase "The spirit is willing but the flesh is weak" into the machine and to translate it into Russian. The translation was produced promptly, but in Russian it read "The liquor is holding out all right, but the meat has spoiled" (157).

Among the problems reported are that "The medical language register in European languages is a jungle of synonyms . . ." (329) and that, therefore, "It may take almost as long to post-edit a MT text as to translate it from scratch" (219). As a result more emphasis is being placed on what is called machine-aided translation where the translator uses automated dictionaries, terminologies, and glossaries to produce copy which must be edited manually.

Those who need translations of scientific articles must, therefore, rely on other sources, such as translation agencies and registries of available translators (86). A survey of a group of British scientists and librarians showed that the steps they took to obtain translations varied with the language. In 67 percent of the French articles, they tried to translate it themselves, while this was true in only 35.7 percent of the German articles and 5 percent of the Russian. The other actions they took were to look for an English summary or abstract (usually a good idea in any case in order to decide whether to pursue a full translation), to try to get a colleague to summarize it, to consult a translator who worked for their agency, to use an outside translator, or to try to locate an existing translation (471).

Existing translations of journal articles are available from several sources. They may be found in "cover-to-cover" translations in which the entire issue of a journal is translated. Journals of this kind, many of which were subsidized by the U.S. government, began to proliferate in the 1950s when it was suspected that Russian science was being neglected by the Western world. By 1967, there were 244 translation journals in various fields of science. Of these, 183 were cover-to-cover. The remaining 61 journals contained translations from more than a single source. Many of the subsidies were dropped when it was realized that scientists do not read even their own journals cover-to-cover and prefer to have articles translated selectively. There are still many cover-to-cover journals in existence. It has been estimated, however, that only 1 percent of all journals are or ever have been available in translation. Those in existence are listed in *Translation Journals,* a list which is published as an annex in the first issue, and in the annual index of *World Transindex; announcing translations in all fields of science and technology,* which is published in Delft as a joint effort of the International Translation Centre and the Centre National de la Recherche Scientifique. Another list which can be consulted is *Journals in Translation,* which was published jointly by the British Library Lending Division and the International Translations Centre in a third edition in 1982 (Boston Spa, England, 1982). It includes 1,077 journal titles, primarily translations into English, as well as a keyword index to titles and an original title index.

Most translations must be obtained as translations of single articles from various depositories in the United States and Europe. The *World Transindex,* published monthly, covers primarily translations from Eastern European (Russian, Polish, Czech, etc.) and Asiatic languages (Japanese, Chinese, etc.) into any of the Western languages including English, French, German, and Spanish. The translations are contributed to the center from different sources including national translation agencies, research institutes, and libraries from several countries. It provides full information on the translation, original article, and the source from which it can be obtained. Each issue contains an author index and a source index of the titles and issues of the journals from which the original articles were obtained. It has been published since 1967, but any issue may include articles published at a much earlier date. The emphasis in each issue is, however, on the literature of the last 2 to 3 years.

A similar publication issued by the National Translation Center in Chicago since 1953 is the *Translation Register Index.* It covers the world literature in the natural, physical, medical, and social sciences, but only those which have been translated into English. Information about book literature in translation is available from the *Index Translationum,* which has been published by UNESCO since 1949 and covers translations of books from all subject areas, into all languages of its member nations.

It is difficult to estimate the probability of finding a translation of a book or article in any of these sources. Bergeijk estimates that only 25 to 30 percent of even the most important literature is translated and that the time between the original publication and the availability of the translation averages 18 months to 2 years (31). It may sometimes seem that the probability of finding a translation of a particular article is as great as that of winning the Irish Sweepstakes. You cannot win in either case, however, without buying a ticket. The investment is considerably less in looking for an existing translation, although the rewards may be great if it can provide a substitute for purchasing an expensive translation *de novo.*

Chapter 7

Writing and Publishing

No piece of research is considered complete until it is adequately reported in the literature.

> Only when he has published his ideas and findings has the scientist made his scientific *contribution,* and only when he has thus made it a part of the public domain of science can he truly lay claim to it as his. For his claim resides only in the recognition accorded by peers in the social system of science through reference to his work. (311:47)

Merton emphasizes that the published report is a *contribution* because the scientist is rarely directly compensated. The research report, Ravetz suggests, is "a piece of property which comes into existence only by being made available to others. . . . A research report hoarded in secret is almost certain to depreciate in value" (377:245). Scientists are, therefore, judged to a large extent on the quality of their writing.

Nevertheless, dissatisfaction with the general quality of scientific writing has been expressed in a number of articles and treatises which call attention to lack of style, absence of clear exposition, and other ills. Some of the failures of communication in science have been attributed to the failure of authors to adequately convey information. Some of the faults of which they are accused are clumsiness of expression, the use of unnecessary jargon and clichés, verbosity, and above all and most damaging, lack of clarity. An anecdote is told about Robert Browning, but also has been attributed to the German poet, Klopstock (416) and probably other poets as well. When Browning was asked by an admiring young lady what an obscure passage in one of his poems meant, he replied, "God and I both knew what it meant once; now God alone knows." A poet may be able to justify such a response, but it is not likely to be accepted from a scientist.

Another source of criticism has been the conventions that appear to eliminate or minimize the involvement of the author in the research process. In affecting an impersonal style, the author seems to be acting as a disinterested observer who is reporting the findings almost as an innocent bystander. As Medawar indicated in responding to his query "Is the scientific paper fraudulent?" (301) this role is to a large extent a scientific fiction. Other observers have suggested that this kind of scientific reporting tends to make it difficult to perceive how the research was carried out (272:28). Among the conventions Gilbert finds in scientific writing are (i) omissions of discussions of difficulties overcome, (ii) exclusion of personal references to opinions and interests of the author, and (iii) use of the passive to eliminate all allusions to the author. "The effect of these devices," he says, "is to emphasize the anonymity of the researcher so that the research becomes everybody's research" (181). Another author expresses the idea this way: "In the name of cultivating an ideal of objectivity, the canons of scientific writing prohibit any reference to self. By strict convention, a scientist may not describe his motives and feelings, his hopes and disappointments, his false starts and errors" (453:ix).

An eloquent disclaimer of impersonal writing in science appeared in a voluminous study for the British government on the comprehensibility of scientific and technical reports. It turned out to be as much a general treatise on scientific language and epistemology as a book on writing. In it the author said:

> The great heresy of science has been its adoption of impersonal language. For every observation involves a private world, and science is based on observation ... we see, then, that to restore the personal pronoun to the language of science is much more than a literary facilitation of readability. It demands the self-understanding of science as an arbitrary enterprise of man. Too long it has masqueraded as the impersonal voice of Nature, another God issuing commandments. And by forcing the restoration we shall force the examination of the values which dominate our choice of axioms. Moreover, this is no mere verbal problem, but a private evaluation of our public human relations. Scientists must restore their own atrophied feelings. (304:13)

Most of the comments on the use of the passive voice in scientific writing, however, have dealt with it not as a cloak of anonymity for the author, but as a rhetorical device which reduces clarity.

Other comments deal with verbosity, awkwardness of style, and the use of jargon. The use of highly technical terms is permissible when you are writing for your peers who can be expected to understand them. It becomes jargon when simple words can convey your meaning more easily. Jargon is a form of verbosity, says Day, which in its extremes can lead to the total omission of one-syllable words (108). This failing was recognized in the early history of science when Thomas Sprat, the historian of the Royal Society, commented in 1667:

> And, in a few words, I daresay; that of all the Studies of men, nothing may be sooner obtain'd, than this vicious abundance of *Phrase,* this trick of *Metaphors,* this volubility of *Tongue,* which makes so great a noise in the World. But I

spend words in vain for the evil is now so inveterate, that it is hard to know whom to *blame,* or where to begin to *reform.*

As a remedy for this condition, he continues, the Fellows of the Royal Society

> . . . have extracted from all their members a close, naked, natural way of speaking; positive expressions, clear senses, a native easiness: bringing all things as near the mathematical plainness as they can: and preferring the language of Artizans, Countrymen, and Merchants, before that, of Wits or Scholars. (417:112)

It would be difficult to find a passage more expressive of the requirements of scientific style. At the same time it makes us aware that style like language, in its complexities of usage and meaning, is as subject to social influences as dress and manners.

Questions have been raised about whether style in the sense applied to literary writing is a useful or necessary consideration in scientific writing. The scientific paper is not the place for an author to exercise flair and style, says Day: ". . . the preparation of a scientific paper has almost nothing to do with writing per se. It is a question of *organization.* A scientific paper is not literature" (108). The conventional scientific paper has developed a format and a style which have been described as proceeding with the stately rhythms of a classical minuet or the regular cadences of an Elizabethan sonnet. However, it differs from good writing of other kinds, says another editor, "perhaps only in requiring a particularly crisp and exact style of writing and in preferring certain conventions and methods of presentation that have come to be regarded as indispensable in science" (332:2). There are no reasons that the principles of good writing are not as valid in scientific papers as in literary essays. In fact, one of the books most widely cited in texts on scientific writing is the little treatise on writing in general by Strunk and White (423). Scientific writing, like all writing, is an effort to engage the attention and interest of the reader. It may even be permissible to use rhetorical devices and stylistic embellishments when they enhance the reader's understanding. One investigator spoke of using an opening sentence which would "grab" the reader, in almost the same sense it would be used by a writer of short stories.

There are a number of excellent texts for scientific authors who wish to improve their writing, including Robert Day's *How to Write and Publish a Scientific Paper,* which appeared in a second edition in 1983 (109), and a collection of essays by two crusaders for clarity in medical writing, Lester King and Charles Roland, published when they were senior editors for the American Medical Association (248). A few additional useful texts are cited in the bibliography (111, 284, 335, 378). Articles on the subject also appear from time to time in scientific journals, some in sprightly style such as William Bean's essays in the *Archives of Internal Medicine* (26) and Richard Asher's Lettsom Lecture published in *Lancet,* in which he describes the emergence of new medical syndromes through the process of being assigned a name (18). Aid can also be found in such standard style guides as the *Style Manual* which is frequently revised by the Council of Biology Editors (91) and the long-enduring *Chicago Manual of Style* (446).

The secret of good writing, professional writers say, is rewriting. Some of the most practiced authors frequently rewrite several times (481). Latour, in his study of *Laboratory Life,* recounts how ". . . advanced drafts pass from office to office being altered constantly, retyped, recorrected, and eventually crushed into the format of this or that journal" (272:49). The key, most advisors agree, is simplicity, the avoidance of unnecessary words, and the elimination of jargon. ". . .[T]he secret of good writing," says one, "is to strip every sentence to its cleanest components" (481:7). No direct relationship has yet been demonstrated between the length of a paper and its significance. Watson and Crick's paper in *Nature* on the structure of DNA was only one page long, yet it "touched off a revolution" (59). Of all the disciplines represented in one study, the life sciences had the shortest articles, with an average length of six pages. This is to some extent a result of editorial pressure, the imposition of page charges, by which some journals require authors to pay publication fees depending on the length of their articles, and the fact that short contributions generally are processed more quickly (59).

O'Connor cites a study which (333:2) listed criteria for effective writing as they had been ranked by 191 successful scientists:

1. Logical rigor
2. Replicability of research techniques
3. Clarity and conciseness of writing style
4. Originality
5. Mathematical precision
6. Coverage of significant published work
7. Compatibility with generally accepted disciplinary ethics
8. Theoretical significance
9. Pertinence to current research in the discipline
10. Applicability to practical or applied problems in the field (75).

Another aspect about which scientific authors are cautioned to take great care is in assigning a title to their work. It may be the most important and informative part of the paper, because it may be the only part a reader sees in an index or in scanning the contents page of a journal. In some information retrieval systems the title becomes the primary retrieval key.

Citation Practices

One of the criteria for effective writing listed above is the citation of the significant published work in the references or bibliography appended to a paper. References serve a number of important purposes. They place the work in the context of previous work and reveal the extent to which the author is familiar with it. They provide the reader with documentation for related research and indicate the author's indebtedness to others. Garfield lists fifteen reasons why an author may cite another article:

1. To pay homage to pioneers in the area in which one is working
2. To give credit for related work (a form of homage to one's peers)
3. To help identify the methodologies, equipment, etc. cited in the paper
4. To refer the reader to related papers
5. To correct something one has reported earlier
6. To correct work reported by others
7. To criticize other work reported
8. To substantiate a particular claim
9. To alert the reader to forthcoming work
10. To refer to work which has not been properly indexed or cited.
11. To authenticate data and facts cited such as physical constants, etc.
12. To identify an original publication in which the idea or concept was discussed
13. To identify the source of an eponymous concept or term
14. To disclaim the work or ideas of others
15. To dispute the priority claims of others (148:189).

Many of these reasons are overlapping, and a single citation may sometimes respond to more than one need.

The kinds of references one should include are listed somewhat differently by Wilson:

1. Those necessary to provide immediate historical background for the problem and previous attempts to solve it
2. Those which provide more complete descriptions of the apparatus or methods used
3. Those from which outside data, facts, equations or arguments used were derived
4. Those which contain either similar or opposing conclusions to those presented.

"Remember," he says, "that it is human nature to magnify one's own contribution and minimize others; therefore, lean over backward in this regard in order to come nearer the correct position on the average" (466:363).

It has been suggested that citation behavior varies from country to country and from discipline to discipline. One study indicated that American psychologists seem to cite primarily the last 5 years of the literature, while their British and French counterparts tend to go back to the nineteenth century. Sociologists are more likely to cite earlier predecessors than psychologists. "The recording of indebtedness may be viewed as an institutionalized form of modesty," says the author of the study, but adds that it is also a way to assign responsibility. While scientific journals may provide information on the form in which references are to be listed, there are, he says, few explicit normative guides to citation practices (239). The relationships between scientific papers established by citations make up a network, says Derek Price. In his analysis he found that scientific papers average about fifteen references each but may range from several hundred to none at all (371). The English crystallographer Bernal, for example, concluded one of his youthful papers with the statement: "I have included no references, as none is necessary" (187:33). Other reasons are sometimes also given, as in a footnote of one author who candidly stated: "Reference omitted to avoid embarrassing its author" (50:71).

References are also sometimes omitted because the information or data have been thoroughly incorporated into common scientific knowledge. Garfield calls this the "obliteration phenomenon." He speculates on what would have happened if Archimedes had written an article on his calculation of the ratio between the diameter and circumference of a circle. First, he says, it would probably be cited as Archimedes' constant and would build up an impressive citation record, until it became so familiar that no one thought it necessary to cite it. This is justified, he admits, but adds:

> But too little attention to citations could be just as dangerous. The conduct of the search for truth requires that assertions be checked, that conclusions be doubted, that results be replicated. . . . Fundamentally, the function of a citation is no different from that of the paper itself: to supply the reader with information he doesn't already have. (164)

Ziman defines the function of the citation in this way:

> The corporate, cooperative nature of scientific argument is made very obvious by the systematic use of *references* or *citations* in scientific papers. It is almost impossible to write or get published without noting explicitly all relevant preceding work by other scholars. A paper without "footnotes" or "bibliography" is suspect; it is an elementary sign of the crank or tyro. (476:58)

Although there are no formal rules for citing or not citing another author's paper, an informal etiquette and ethical understanding have grown up around the practice. It is considered bad manners, if not unethical, not to cite an author when it is appropriate to do so. On the other hand, to cite a paper which is not relevant or worse which has not been read or has not been understood because it is in a foreign language is regarded as "amoral" (208:83). Chargaff thinks this has now become a common practice:

> Bibliographies were comparatively honest, whereas now entire packages of references are being lifted by a form of transduction, as it were, from one paper to the next; so that if some work gets into the habit of not being quoted, it never will be so again. (74)

An author may fail to cite a work because of unawareness of its existence, because it is unattainable, or because it is in a language that is not understood. Such omissions are nevertheless frequently regarded as conscious oversights or even attacks by the authors who are not cited. Wade speaks of Schally's sense of "outrage" when his rival Guillemin omitted a reference to some earlier work of his which he thought was relevant (453:122). Failure of this kind, real or imaginary, is regarded as widespread. One sociologist reported that 50 percent or more of American scientists believe that others have failed to properly refer to their work (177:13). He adds:

> Concern about citation to previously published research is prevalent in the scientific community. If any scientist doubts this, just remember what a colleague had to say the time he sees a paper that should have cited some of his publications. This failure is a frequent occurrence and is often seen as an intentional act by certain individuals who do not want to give credit where credit is due. (177:60)

Waksman makes a similar charge:

> . . . one can easily pick out of any issue of the *Journal of Immunology* or other immunologic journals at the same level a half dozen papers in which the principal work in the field, often work anticipatory or even duplicating that which is the subject of the paper, is not referred to. Whether the reason for omission (author's ignorance, purposeful obfuscation) there is obvious failure of the refereeing function here. (454)

Waksman places the responsibility on the journal editor, but it is clearly that of the author who has a moral obligation to be aware of and fully acknowledge the work of predecessors. Failure to acknowledge the source of an idea is, Garfield says, a form of plagiarism, not the kind in which complete texts are copied, which he regards as "grand larceny," but where only the ideas are used, a form of plagiarism more difficult to detect (159). There is a good reason, says Wilson, for a "generous attitude" with regard to references:

> Scientists form one group which is practically never deceived by men who push themselves forward on the work of others. Failure to give proper credit to another's work can generate more bitterness than any other action. . . . Few secrets are hidden, and a man who infringes on the intellectual property of others will have his sins passed around the world with amazing rapidity by the gossip of his fellow scientists. Careers have been damaged for life by a few thoughtless acts of this kind. Half a century of righteous behavior may not be sufficient atonement. (466:363)

Correct Citations

Incorrect citations are not as much an issue of ethics and good manners as they are failures to meet an obligation to the reader. They not only create unnecessary work for the reader, but reflect discredit on the author and may raise questions about the validity of the rest of the work. Yet an experienced editor says ". . . there are far more mistakes in the literature cited section of a paper than anywhere else" (109:38). One study was based on a sample of thirty-four "significant journals of science" from which three articles were selected at random from a single issue and the references were checked for accuracy. The investigators found 466 errors in 367 references out of a total of 2,448 references derived from 102 articles, or an average of almost four per article. The bulk of the errors occurred in the authors' names (over 75 percent). The other errors were fairly equally distributed among volume numbers, pagination, and year of publication (364). Another study followed a single journal over a 13-month period and found that 54 percent of the references in the accepted articles had one or more errors that needed correcting (244).

Errors in literature references tend to occur in patterns, as a result of faulty transcriptions, transposed numerals or letters, or misinterpretations of hand-written notes or oral transmissions (261). There is hearsay evidence that John Hunter, who was a baronet, was once listed in a bibliography as "Bart, John Hunter," but an actual case has been reported in which the author Martin Rushton, whose name appeared at the head of his article with the qualification "M.A.Cantab," appeared in the bibliography as Rushton and Cantab (283). Errors in references tend to be perpetuated. The classical story involves a Czech paper on dysentery which was frequently quoted and attributed to the author O. Uplavici. He appeared in many reference lists and even in a biographical directory. There is an apocryphal story that he was once considered for an award. He was finally laid to rest by the parasitologist C. Dobell, who discovered that "O. Uplavici" meant "On dysentery" and the original article had been written by J. Hlava (116).

In addition to the paramount need for accuracy in citing references, the scientific author is also faced with the requirement to cite them in a form which is acceptable to the editor. At a meeting of life sciences editors in 1977, it was reported that thirty-three different styles of citing were found in fifty-two journals. Further, it was estimated that given the differences in abbreviations, formats, styles, order of the citation elements, and all the possible combinations and permutations there were 2,632 possible variations (334). O'Connor has formulated a variation of Murphy's Law which states ". . . if an editor can reject your paper he will; and its corollary: if you submit the paper to a second editor his journal invariably demands an entirely different reference system" (334). It is wise when submitting a paper to follow carefully the instructions to contributors which most editors insert at the beginning of each issue of the journal.

There are essentially three general styles of citing references:

1. The so-called Harvard system, in which the names of the authors and the year of publication are given in the text, and the full citation is listed in an alphabetically ordered list of references

2. The sequential numbering system, in which the references are listed and numbered in the order they are cited in the text
3. The alphabetical-numerical system, in which the references are listed in alphabetical order and cited by number in the text (109:43).

Sequentially numbered systems are said to have the advantage that the reader can more easily locate that portion in the text in which a particular author is cited (151). However, they make it more difficult to interpolate new citations when they are required in revising a paper.

There have been numerous pleas for a uniform style of citing (422), and efforts have been made to standardize citation styles. A large number of editors convening in Vancouver in 1977 agreed to accept a uniform style for citing references in all papers submitted to them, although they still retained the right to modify the style to conform to their own policies. These standards have been widely disseminated in the biomedical literature (228). The standards recommend that all references be numbered consecutively in the order they are first cited. The citation number is repeated if the same article is cited again later in the article. Other recommendations relate to the number of authors per paper to cite (all six when there are six or fewer, but only the first three when there are more than six, adding *et al.* for the others), the order of the citation elements, and the abbreviations to be used. Efforts have also been made to standardize abbreviations for journal titles which have been accepted by the major abstracting and indexing services as well as biomedical editors. Standard abbreviations can be found in such publications as the *List of Journals Indexed in Index Medicus* published each year by the National Library of Medicine, the *Bibliographic Guide for Editors and Authors* published under the auspices of the American Chemical Society, the Bioscience Information Service of Biological Abstracts, and the Engineering Index, as well as the *Science Citation Index Guide and List of Source Publications* issued each year by the Institute for Scientific Information.

Authorship

Another problem relating to the ethics and etiquette of scientific papers is who should be listed as authors and in what sequence. The first author on any paper receives most of the citation credit, although all authors add another item to their bibliographies. This was not much of a problem when most papers were produced by single authors, but there has been a considerable rise of multiple authors in the past several decades. This has been a result to some extent of the increase in the complexity of modern science and the growth of interdisciplinary research. In the field of high-energy physics, for instance, it is not unusual to find articles with thirty or forty authors (162). One paper on nuclear physics published recently had fifty-eight authors and was the result of a collaboration among scientists from four countries (37). While this is rare in the life sciences, there are exceptions. One paper in *Transplantation Proceedings* in 1971 listed thirty-six authors (250). Nevertheless, less than 1 percent of all the articles indexed have more than ten authors (432). The number of articles with single authors in the *New England Journal of Medicine* and its predecessor the *Boston Medical and Surgical Journal* dropped from 98.5 percent in 1886 to 4.0 percent in 1977. The most dramatic change took place between 1946 and 1976, when the percentage of single authors decreased from 49.0 to 4.0 (121). The average number of authors for scientific papers in general rose from 1.67 to 2.58 between 1960 and 1980 (59). The average number of authors for biomedical papers in 1963 was 2.26 (77). These data can be compared with a list of the 300 most cited papers published between 1961 and 1975, in which only 12 percent had a single author and 27 percent had four or more (168).

The question of who to list as an author and in what sequence can become, says Dudley, a vexing problem which can "rend apart a previously close-knit department" (120:19). The author who is assigned the primary position is generally regarded as having played the major role in performing the research and preparing the report, although this is not always the case. It has been indicated that there are cases in which institutional directors and departmental heads may list themselves as authors even though they have not been involved in the research. Chernin compares this practice with the custom of "the medieval seigneur, whose prerogative it was to spend the marriage night with each new bride in his domain" (76).

There do not seem to be any generally accepted guidelines although the issue has been discussed at length in the literature. Some of the conventions derive from history and national customs, such as the old practice of naming the dissertation director as *praeses* on a thesis with the actual author. Two investigators who frequently publish together often find themselves changing places as first authors. One journal adopted a practice of listing authors in alphabetic order to solve the problem of assigning priority. After following this policy for a while, however, the journal found it was getting fewer contributions from authors whose names began with P through Z (162). It may sometimes not matter whose name goes first. In an Institute for Scientific Information study of the 1,000 most cited authors of journal articles from 1963 to 1978, it was discovered that eminent scientists received twice as many citations as secondary authors as they did as primary authors. They also published more papers as secondary authors than they did as primary authors (162). This phenomenon is reflected in the remark that is attributed to the Nobel Prize winner Schally when the issue was raised: "What did it matter to me?," he said. "It's my lab— I get the glory anyway" (452).

There is general agreement that only those who have been significantly involved in the research and writing should be listed as authors. This decision may be difficult in projects which have involved dozens of workers including basic scientists, clinicians, statisticians, and laboratory workers in various stages of designing the protocols, carrying out the investigation, and analyzing the results (36). Among the questions which have been suggested for consideration are:

1. Who was responsible for the research idea?
2. Who designed the methodologies?
3. Who wrote the report?
4. Who did most of the physical work?
5. Should the statistical consultant be included?
6. Should salaried research assistants be included? (138)

It is generally agreed that assignment of authorship must be accompanied by some responsibility for the contents of the report. Some of the co-authors in the flagrant cases of fraud reported earlier were not aware that their collaborators had submitted false data. In the case of the young Harvard investigator who co-authored papers with senior members of the faculty who did not know he had falsified his findings, a review panel indicated that they shared the responsibility. Commenting on this case Relman states categorically:

> Anyone who allows his name to appear among the authors of a paper assumes major responsibilities. Even if co-authors have not actually done any of the laboratory work, they should at least know that the experiments and measurements were carried out as described, and they ought to understand what was done and why. Co-authors should be able and willing to defend the paper in public, and that means they must be confident about the integrity of the data. Furthermore, co-authorship should never be conferred or accepted as an honor or simply as a reward for providing resources or sponsoring a junior investigator who has done all the work. (380)

Getting Published

One of the author's major decisions in publishing a paper is choosing the journal to which the manuscript is submitted. The author will naturally wish to choose the one with the most prestige, the largest circulation for the targeted audience, and the shortest publication lag. The author will also wish to minimize the chances of being rejected and going through the long submission cycle again, a process which may require reformatting the paper. The important elements in choosing a journal, according to McDonald, are relevance to the author's interest (scope, readership), prestige, quality (appearance, readability, etc.), refereeing competence and fairness, publication lag, and last (although it probably should be listed first), probability of acceptance (246:63).

Size of circulation is not always associated with prestige, although it does expose the work to a wider audience, if that is the author's goal. Crick referred to the *Proceedings of the Royal Society,* which published the detailed account of the structure of the DNA molecule, as "an obscure journal" and chose to make the first announcement in the large-circulation journal *Nature* (96). The circulation of a particular journal may sometimes be determined by looking for the "statement of ownership, management and circulation" required by the U.S. Postal Service which appears once a year, usually in a November issue. Other sources are such guides as *Ulrich's International Periodical Directory* (443) or the *Directory of Publishing Opportunities in Journals and Periodicals* (115), both of which provide circulation data for many foreign as well as U.S. titles.

The "prestige" of a journal is more difficult to determine and may tend to wax and wane with the times and changes in editorial responsibility. The National Academy of Sciences survey on the life sciences defines prestige this way: "In almost every scientific field there is a hierarchy of journals that reflects the relative quality of published papers. Although it does not exist overtly, this hierarchy is known to all sophisticated scientists within the field" (318:413). One of the best ways to determine the prestige of a journal, then, is to consult your colleagues. A more objective method is provided by the *Journal Citation Reports* published by the Institute for Scientific Information, which calculates an "impact factor" for journals based on the number of citations in a given year to papers in a particular journal divided by the number of "citable" papers in that period (150:149).

Publication lag, the time between submission of a manuscript and its appearance in print, also tends to vary considerably among journals. Delays of a year or more are not uncommon. They depend to some extent on the frequency of publication of the journal and the backlog of manuscripts it has accumulated. Estimates of time lag may be made in some cases by comparing the dates "received for publication" which appear on some published articles with the cover date of that particular issue. Journals tend to be divided into rapid publication journals like the "letters" journals (*Tetrahedron Letters,* etc.), wide-distribution journals usually published weekly like *Lancet, Science, Nature,* and the *New England Journal of Medicine,* and the specialty journals which usually appear monthly or quarterly. The editors of *Science* promised to notify authors in 4 to 5 weeks in accepting or rejecting short articles and 6 to 10 weeks for longer articles (226). The editor of the *New England Journal of Medicine* stated in 1974 that: "The interval between the receipt of a manuscript that is eventually published and notification of authors of acceptance average 33 days." He added, however, that some reviewers may take as long as 32 weeks in returning a manuscript and did not comment on the interval between acceptance and publication (223). The average lag time between submission and publication seems to vary from one discipline to another. In one study it was 5.5 months for physics, 6.2 months for chemistry, 13.2 months for political science, and 13.6 months for sociology (40). In other studies it was 5 months on an average in medicine and 9 months in psychology. One study selected an article in microbiology from each of fifty-one journals which publish in that subject. In about half of the articles it was possible to compare the receipt and acceptance dates with the publication date. The mean elapsed time was 35 weeks, with a range of 15 to 89 weeks (72).

Authorities emphasize that the selection of the journal in which you wish to publish should be an early step in planning a paper. Journals even in the same discipline differ in scope and in kinds of topics and formats they accept (220). Unless you are familiar with the publication, it is wise to scan several issues or to consult the title pages of the journal in successive issues of *Current Contents.* In many journals the editorial policies are printed with the instructions to prospective authors. A compilation of these editorial guidelines for 246 English language biomedical journals was published in 1980 which, in addition to the editorial and reference guides for these journals, includes the uniform requirements for manuscripts cited earlier (303).

There are other concerns about submitting manuscripts which involve ethics and etiquette. One controversy involves whether to report work in progress. Reporting work before it is completed, says Durack, is a result of two forces, one which seeks to increase the number of publications to one's credit by reporting the same data several times, and the other a result of the fear of prior announcement by someone else working on the subject (121). An editorial in *Nature* denounced the first practice as adding unnecessarily to the bulk of the literature and as "creating bibliographic chaos" by establishing an uncertainty about which of the author's articles to cite. It also characterized "preliminary publications" which do not result in full versions as "bad author behavior." The editorial goes on to criticize papers which, once in print, also appear in published symposia and conferences (375).

There may be conditions under which preliminary announcements can be justified such as when research has reached a point where it can be shared with a wider audience or to identify yourself as a co-worker in a particular field (424:45). Even inconclusive results can be useful, as Benjamin Franklin indicated in a letter on his kite experiments that he wrote in 1753 to his London friend Peter Collinson, who served as his liaison with the Royal Society:

> These thoughts, my dear friend, are many of them crude and hasty, and if I were merely ambitious of acquiring some reputation in philosophy, I ought to keep them by me, 'till corrected and improved by time and further experience. But since even short hints, and imperfect experiments in any new branch of science, being communicated, have often times a good effect, in exciting the attention of the ingenious to the subject, and so becoming the occasion of more exact disquisition and more complete discoveries, you are at liberty to communicate this paper to whom you please, it being of more importance that knowledge should increase, than that your friend should be thought an accurate philosopher. (210:108)

Simultaneous submission of a manuscript to more than one journal, on the other hand, is regarded as highly unethical. The editorial in *Nature* referred to above cites a case in which an author submitted an article to three journals at the same time. It was rejected by one, but it was accepted by two. Besides being embarrassing to the author, it created unnecessary work on the part of two of the journals. This behavior is sometimes justified on the basis that some journals are extremely slow in responding or are likely to reject the paper. In the second instance it reflects poor judgment on the part of the author. In the first case, a better procedure, suggests an experienced editor, is that, if after a reasonable length of time a response is not received, the author should notify the editor that it is being withdrawn before offering it to another journal (112). Some journals such as the *New England Journal of Medicine* have strict rules about any prior disclosure of any kind, and will not accept an article if the research results have been reported even in a newspaper (223).

Editorial and Review Process

Editors of scientific journals have been called the "gatekeepers" of scientific information (113). "The editor's task," as defined by the National Academy of Sciences survey of the life sciences, "is to decline work that is duplicative, incompetent, incorrect, or totally pedestrian" (318:413). Functioning sometimes alone, sometimes with an editorial board, and sometimes with a few or with many reviewers, the editor has the primary responsibility for determining the scope and maintaining the quality of a publication.

These practices have a history going back to the early days of scientific associations. All papers submitted to the *Académie des Sciences* in Paris in the seventeenth century were reviewed by appointed members of the academy before they were permitted to be published under its auspices. The *Philosophical Transactions* in the seventeenth century were a private undertaking of Oldenburg, although he was at the time a secretary of the Royal Society. It was his responsibility to select items for publication from the correspondence and other information that came his way. The *Transactions* in the seventeenth century, therefore, contain some bizarre and incredible accounts as well as the important observations of Boyle, Hooke, Leeuwenhoek, Newton, and others. When a paper such as Newton's important communication on his experiments with the refraction of light was presented to the Society itself, it was received with enthusiasm but nevertheless was "ordered" to be reviewed by several of the Fellows before it was formally received (202). In 1752 when the Royal Society took over formal responsibility for the *Transactions,* the Society appointed a "Committee on Papers" whose task it was to select those papers which would be published (260:136). This is sometimes described as the origin of the peer review system in scientific journals, but other precedents are to be found. In a French translation in 1740 of the *Medical Essays, Revised and Published by a Society in Edinburgh,* the process is described almost as it exists today:

> Memoirs sent by correspondence are distributed according to the subject matter to those members who are most versed in these matters. The report of their identity is not made known to the author. Nothing is printed in this review which is not stamped with the mark of utility and if the number of good memoirs is not sufficient to make a paper volume, they wait for a new harvest until it can be achieved. (131)

Although the practice of soliciting the opinions of experts (peer reviewers) before deciding to accept, return for revision, or reject a manuscript is widespread among scientific journals, it is by no means universal. The *American Journal of*

Medicine, an outstanding clinical journal, used no outside reviewers when it started. As a result, it was said to have reduced the lag time in accepting a paper to 2 weeks or less. On one of the leading British medical weeklies, *Lancet,* the decision is made by the editor for about 90 percent of the papers published (223). A survey of 156 journals in the physical and biological sciences published in thirteen countries found that only 71 percent made some use of outside reviewers (referees). The low percentage was attributed to the fact that the editorial staff of some European journals has the responsibility for acceptance or rejection (299:38).

The common peer review process is described by the editor of the *New England Journal of Medicine*: Each manuscript needing review is sent to at least two referees who send their appraisal of the overall acceptability and merit of the contribution to the editor, who may send a summary of the criticisms and comments about the substance and the style of the paper to the author. The editor makes the decision on whether or not to publish the paper, but additional opinions may be solicited if thought warranted (379:75). A comprehensive survey of the review process in scientific journals found that two referees per paper was the average, although it tended to be more for longer papers and papers that were eventually rejected (237). Referees are usually not paid, and very few receive acknowledgment of their services in the journal (237). The journal *Science* claims to have over 6,000 reviewers listed in its files that it can call on (2), a number which seems reasonable considering the variety of disciplines covered by the journal.

Despite the numerous criticisms of the peer review system, there seems to be little empirical evidence on how it functions (467). The journal *Behavioral and Brain Sciences* in 1982 published an experimental study of the review process. In its customary fashion it solicited and published the comments of fifty reviewers, some of whom commented on the study's methodological weaknesses and the ethical questions it raised (as a necessary consequence of the research design the investigators could not obtain the informed consent of the subjects). They selected one article from each of twelve psychological journals which had been published up to 3 years earlier, modified the title and opening sentences, gave fictitious names to the authors and their institutional affiliations, and resubmitted them to the journal which had published them originally. Only three of the papers were detected as having been previously published, only one was accepted for republication, and eight were rejected. The fact that the genuine authors whose papers were accepted were associated with institutions of high reputation seemed to point clearly to reviewer bias (357), at least in the case of journals in psychology.

Testimony to this effect was offered by one of the commentators who reported that some articles he had submitted to "mainstream psychological journals" when he was a junior member of a midwestern university were rejected, but were published in those journals after he joined the faculty of Harvard University (385). Another commentator readily admitted that as an editor he could guarantee that an article would be rejected by carefully selecting the reviewers (35). A recently initiated virology journal recognized this as a factor by announcing that authors would be permitted to indicate one individual who should be excluded from reviewing the manuscript (290). While the possibility of reviewer bias is recognized, it is not uniformly regarded as a significant problem in the peer review process. The study by Zuckerman and Merton, for instance, found that, at least in physics, it was not necessarily the more established scientists who were more readily accepted for publication and that in fact the younger scientists tended to have papers accepted at a greater rate (484).

It has been suggested that bias might be reduced by maintaining the anonymity of the authors so that reviewers would not be influenced by their reputation or institutional affiliation and by making the reviewers' names public so that they would have to defend their judgments. Not disclosing the author's name is apparently not a widespread practice. In a survey of fifty journals published by national associations in seven disciplines including the life sciences, only nine journals maintained the anonymity of the author in the review process, and eight of these were in sociology (93). Although he argues for maintaining the anonymity of journal referees, Shils comments that this is not true in law or in sports, where referees are highly visible (404). On the other hand, authors are seldom told the names of their referees. Reviewers' identities are usually concealed not only prior to the acceptance or rejection of a manuscript but in most cases even after the papers are published and are not even made known to the other referees assigned to the manuscript (237). Referee anonymity can provide a "tempting invitation to be irresponsible, to be excessively dogmatic or critical, to be careless. . . ," says one author (155). On the other hand it has been argued that anonymity protects less powerful reviewers from possible retribution and that identifying reviewers might otherwise compromise the effectiveness of the review process (358). Reviewer anonymity has been strongly defended:

> Some reviewers feel uncomfortable with the apparent power that anonymity gives them to judge in secret. However, reviewers only recommend to editors; and editors are not anonymous. The reason why anonymity is required is that it allows a level of honesty and candor that simply could not be achieved in any other way. . . . Anonymity minimizes polarization. (407)

The peer review process has also been criticized on other grounds. The increase in the number of papers submitted, says Waksman, has resulted in an increase in the number of inexperienced referees who are often at a level of expertise the same as or lower than that of the author. Referees, he adds, sometimes in their zeal to find something to say about a paper may provide inept or capricious criticisms. The process, he says, causes inordinate delays in disseminating scientific information through the process of rejection and resubmission (454). Visscher regards refereeing as a form of censorship, as indeed must be the case whenever literature is rejected for one reason or another. "Editors of journals," he says, "set themselves up as watchdogs of quality, but frequently do not have either the competence or the time to fulfill their functions in this regard" (451).

McCutchen points out cases in which failures or errors in referee judgments have kept important work from being published or delayed it considerably. Reviewing, he says, is a means for keeping the rate of change in science down to a level acceptable to the establishment, which is generally conservative. "The review system," he concludes, "is poisoning the atmosphere in science" (286). Stumpf lists other shortcomings in the process:

1. For the most advanced scientists only a few or no peers exist.
2. The closest scientific peer is a competitor.
3. The process provides no opportunities for rebuttal.
4. Anonymous reviewers cannot be held accountable.

He recommends a system of signed reviews with opportunities for rebuttal by the author (424). Wilson argues further that the review process serves in many cases only to delay publication. Of all the papers submitted in one year to the *Journal of Clinical Investigation,* he reports, 85 percent of those rejected were published elsewhere, some of them not until 3 to 5 years later. Most of these were "not changed or changed only in minor ways." The peer review process, he concludes, had a significant impact on only one-third of the papers submitted, the 17 percent which were revised before resubmission and the 15 percent which were not published (467).

Another source of criticism derives from observations on the concurrence of judgments between pairs of reviewers. One author insists:

> In a typical journal one quarter of the papers meet with two favorable reports, one quarter are reduced to ashes by both reviewers, and one-half receive contradictory evaluations (330).

There is some evidence that tends to support this view, although consistency in reviewer judgments seems to relate to "the degree of quantification" in a particular discipline, or what has been called the level of codification (483:507) (see p. 9). In one study there were only five disagreements in reviewing 172 papers in physics, a concurrence rate of 97 percent, whereas the level of agreement in reviewing 1,572 papers submitted to biomedical journals was only 75 percent (299:38). According to Ingelfinger, who studied the concurrence rate between two reviewers for 500 papers submitted to the *New England Journal of Medicine,* there was an ever wider range of divergence. In fact, he concluded that the results were only slightly better than chance. The concurrence on the accepted papers (20 percent of those submitted) was 50.5 percent, but only 39.7 percent for the rejected papers, with an average of 41.8 percent for all the papers (223).

Acceptance and Rejection

Every author who submits a manuscript to an editor for publication must be resigned to the anxiety of waiting to see whether it has been accepted, returned for suggested revisions, or rejected. If it is rejected, the author is in good company. A survey of authors in all fields of science and technology in 1977 revealed that 44 percent of all articles were rejected by the first journal to which they were submitted. Of these, 28 percent were resubmitted and published by another journal (246:68). There is little comfort in knowing that some distinguished scientists in the past have had the same experience. Even Nobel Prize winners have not been spared. Hans Krebs' first paper on the citric acid cycle in 1937 was rejected by *Nature*. It was returned to him, he says in his *Reminiscences and Reflections,* "five days later accompanied by a letter of rejection written in the formal style of those days" (257:99). The full paper was published in *Enzymologia* 2 months later. Krebs adds:

> If one glances today at the *Nature* of 1937 and examines the kind of papers which the editor felt ought to be printed, one is struck by the difficulties that exist in deciding what is worth publishing. Most of the material published in the form of letters has turned out to be of no particular importance. (257:99)

Another of the charges against the refereeing system is that it is resistant to innovative and unconventional ideas which do not fit the current scientific "paradigms." Kuhn has commented that ". . . unexpected novelties of fact and theory have characteristically been resisted and have often been rejected by many of the most creative members of the professional scientific community" (265). Horrobin echoes this view, saying: "History has repeatedly shown that new research which is genuinely important is often vehemently rejected by established scientific opinion" (218). It is also true that papers even by Nobel Prize winners can be turned down not because they were too innovative but because they allegedly lacked novelty. Thus, one of Guillemin's papers was rejected by *Science* in 1969 with the statement that "it was well known that vasopression releases TSH in *vitro* and in *vivo*" (272:117).

There are, nevertheless, few documented cases of actual suppression of scientific papers. Even Galileo found a way of making his "heretical" views known. The last paper that John Hunter submitted to the *Philosophical Transactions* in 1792 was his "Observations on the Fossil Bones." In it he suggested that the earth might have existed beyond the appointed biblical span. The paper was rejected after Hunter refused to remove the statement that the fossil bones might have existed for many thousands of years. It did not get published until 68 years later when it appeared as a part of his collected works (252:298). The editorial process in the late eighteenth century is also revealed by the fate of Edward Jenner's first paper on his experiments with inoculations against smallpox, which he sent to the Royal Society in 1796. It was referred for review to Everard Home, who recommended rejection on the basis that it represented only a single case and suggested that Jenner try the procedure on at least twenty or thirty children. At this point, he said (perhaps with justice): "I want faith." The President of the Royal Society sent it back to Jenner without referring it to the Publications Committee. Jenner's *An Inquiry into the Causes and Effects of the Variolae Vaccinia* was not published until 2 years later and then at his own expense (48).

One of the reasons that scientific ideas eventually reach their audience is that a majority of the rejected papers are resubmitted to other journals. Of all the papers rejected in 1970 by the *Journal of Clinical Investigation,* which usually rejects about half of the manuscripts submitted, 85 percent were subsequently published elsewhere—3 percent in the same year, 16 percent in the following year, 37 percent in the third year, and some as much as 6 years later. Many of these were published in journals of equal prestige (467). Waksman cites a study of a journal in 1979 which had a rejection rate of 40 percent. Two-thirds of the rejected papers were promptly resubmitted, and 90 percent of these were ultimately published, "often in a journal of equal quality to the first" (454:1011). In psychology it has been estimated that about one-fifth of the articles which appear in the principal journals have been rejected at least once. The comment was added that "a persistent author will seldom fail to get his manuscript published eventually" (75:358).

Resubmission rates, like acceptance–rejection rates, seem to vary widely according to discipline. Thus, the 1977 survey indicated that the lowest resubmission rate was 10 percent in physics, which also had the lowest rejection rate, 19 percent. The highest resubmission rate was 44 percent in psychology. The life sciences were also at the higher range with 33 percent. The rejection rate was 71 percent in psychology and 48 percent in the life sciences, so there is apparently some correlation (246:68). The rejection rates go even higher when one ventures into the humanistic disciplines. Another study found that the rejection rate in the humanities in general was 75 percent as compared to an average of 25 percent in science. A suggestion was made that this might also be a function of the higher levels of support for scientific publication as well as the fact that scientific papers usually go through a more rigorous preliminary review in formulation of the research project and in exchanges with colleagues. The rejection rates ranged from 90 percent for historical journals to 20 percent for journals in linguistics, with 29 percent for the biological sciences and 24 percent for physics (484).

There are, of course, other factors besides the discipline which are reflected in acceptance–rejection rates. Two journals in the same discipline may have widely divergent rates. Adair reports that *Physical Review,* an important physics journal, has an acceptance rate of 80 percent, which is characteristic of the discipline, while a companion journal, *Physical Review Letters,* has a rate of 45 percent because only those papers that "are exceptionally novel and newsworthy are published" (5). "High-prestige journals," says Huth, "also have high rejection rates" (220:8). There also seems to be a correlation between rejection rates and the size of the journal's circulation. *Science,* with a circulation of about 150,000, rejects about 80 percent of the manuscripts submitted. The *New England Journal of Medicine*'s circulation is over 200,000, and its rejection rate is 85 percent (222). In his study of refereeing practices, Juhasz found that on the average large scientific–technical journals accept 1.2 articles for every one they reject, while small journals accept 2.3 papers (237:182).

The peer review system used by scientific journals has been, as we have seen, subjected to considerable criticism, and suggestions have been made to improve it (237). The general consensus, however, remains that, despite its shortcomings and dangers, peer review provides an essential quality filter for the literature and that it is "less dangerous than the alternatives" (434). Studies have repeatedly shown that a small core of journals bear the major burden of disseminating scientific information. This would not be possible without some sort of a qualitative filtering process such as the peer review process provides.

Chapter 8

Indexing Languages

Indexing as a means of controlling and gaining access to the literature has a long history. The first volume of the *Philosophical Transactions* completed in 1666 contained an index of titles "abbreviated in an Alphabetical Table" and followed by an index of the contents "digested into a more natural method." The alphabetical table was based on keywords in the titles and in the articles. The "more natural method" consisted of a kind of classified arrangement of topics. It apparently was not successful because it was not repeated in the second volume. The first separate periodical index was issued by an obscure Flemish bookseller in 1683 under the title *La France Sçavante* as a guide to the volumes of the *Journal des Sçavans* which had appeared up to that time. It consisted of three parts or "conspecti," a register of the articles as they appeared *(chronologicus)*, an author index *(personalis)*, and a classified subject index *(realis)*. It is interesting to note that this was the same tripartite division adopted by the *Current List of Medical Literature* in 1950 when the National Library of Medicine decided to upgrade a title list which had been used during World War II as a current awareness medium for military medical officers. Some of the current methods of organizing and controlling the literature can be regarded as modifications and variations of these early attempts to develop indexing techniques and languages.

Anyone who has ever tried to put together a subject index to a personal file knows how difficult and intellectually challenging a job it can be. Access to ideas, concepts, and named objects is achieved through words. Language provides the tenuous web which holds all our knowledge together as well as the means by which we communicate with one another. Yet to some extent all languages are private, filtered through each individual's own experience. The organizing principles are similar in both personal and shared information systems. The problems, however, are compounded with shared systems because the individual who stores the information is usually not the same as the one who retrieves it. These systems must, therefore, be based on indexing languages which

109

are also shared. There are different forms of indexing languages, but like all languages they are all composed of two parts: a vocabulary, a set of terms on which there is a reasonable consensus as to meaning, and a syntax, a set of conventions or rules for using these terms. Like all living languages indexing languages are, moreover, not fixed or immutable, but change as new concepts are introduced and as new relationships are recognized.

New words are constantly being introduced in the life sciences and discussed and debated in the literature. Because of the dependence on words in the communication of scientific ideas and the organizing of scientific information, much effort is spent by scientific organizations to establish a consensus about the terms used in scientific discourse. Many scientific disciplines are represented by international committees and commissions which have been engaged for a long time in attempting to establish officially recognized terminologies. This is particularly true in the so-called taxonomic sciences where genera and species are constantly being discovered, reevaluated, or redefined on the basis of a clearer understanding of their morphology and composition. The International Commission on Zoological Nomenclature has been functioning since 1895 (44:392). There are many other similar bodies concerned with nomenclature and terminological problems, both in the taxonomic sciences and in such areas as nomenclature of drugs and disease.

In spite of these efforts to control terminology, the language as it is used in scientific writing presents many problems to indexing systems which are trying to maintain the kind of logic, consistency, and uniformity which would make them easier to use. Ambiguities, for instance, are introduced by the presence of a large number of synonyms for some terms. One authority states:

> Between one quarter and one half of medical and paramedical terminology — depending on the specialty — consists of obsolete, rare, officially rejected or otherwise undesirable synonyms. The number of such undesirable synonyms runs into tens of thousands.

As an example, the author cites the yeastlike fungus called *Candida albicans,* which he found in the literature under 170 different names (293).

The problem is compounded by the fact that few of these synonyms have identical meanings. People may also use the same term with different meanings, although this is perhaps less true in science than in politics. Organisms can have both common and scientific names, and chemical compounds and drugs can have popular, proprietary, systematic, or code names. Homographs, words which are spelled alike but have different meanings, *crane* (bird) and *crane* (tool), also add ambiguity. Homonyms, words which sound alike but are spelled differently, *right* and *write,* do not provide any problems in indexing systems since they are listed separately. There is another class of terms called polysemous, in which the homograph has related meanings which differ according to context. The term *plasma* has a different connotation in biology than it does in physics, for example. Antonyms, terms with opposite meanings, also play a role in indexing languages, because the indexes may only have an option to use one term, say *roughness* instead of its antonym *smoothness* to cover references to variations in these qualities. Another problem encountered in indexing language is that the same word can appear in a number of different forms: as the singular or plural, as an abbreviation, as an adjective or a noun, or as a spelling variant, such as the English spelling *foetus* or the American spelling *fetus*. Indexing languages may also have different conventions for dealing with compound words such as *Kidney neoplasms* by separating them (as they would in post-coordinated languages) or combining or inverting them (in pre-coordinated languages). Another characteristic which affects indexing languages is the generic–specific relationships between terms, which we will turn our attention to when we discuss classification as a primary method of organizing information.

There are essentially two kinds of indexing languages, those with uncontrolled vocabularies, or "natural language" systems which use the words that appear in the author's title or text, and those with controlled vocabularies in which the terms that can be used for indexing are selected in advance and rigorous attempts are made to eliminate synonyms and to conform to rigid standards of nomenclature and usage. Actually, these descriptions represent two extremes in a continuum of forms of indexing languages, because none of the indexes we shall examine is a pure example of either. All natural language systems have words such as "and," "the," etc., which are excluded from their indexing vocabularies. They also have highly developed conventions and auxiliary approaches to compensate for some of the deficiencies of natural language systems. Many controlled vocabulary systems tend to use natural language, that is, familiar terms rather than technical terms, and to respond on a continuing basis to changes in the language as they occur. They all provide terms, called tags, descriptors, or subject headings, intended to lead an investigator to the information being sought. This happy conclusion can only occur if the investigator learns to understand and to use the language of the system being used.

Natural Language Vocabularies

One of the problems with controlled vocabulary systems is that they require human indexers to select the appropriate terms to describe what they think are the significant aspects of an article. The process tends to increase the costs required to maintain the system and to lengthen the lag between the time an article is received and is represented in an index. Controlled vocabularies, moreover, are more expensive to develop and maintain. It has been estimated (455) that it cost as much as a million dollars to produce one thesaurus of terms developed for the Department of Defense (433). Even a small-scale thesaurus of 2,500 terms may take the full time of two individuals for a 6-month period (415:13). Controlled vocabularies also require the user to be more familiar with the terms and how they are used. For this reason, increasing attention has been paid to machine methods as computers became available for analyzing, sorting, and arranging title and text words as they appeared in the articles.

With the growth of machine-readable databases, uncontrolled or natural language indexing systems came into much wider use. They are based on the ability of the computer to rapidly permute words in an author's title and to display them alphabetically as index terms in relationship to the other title words. The technique has been called Keyword in Context or KWIC indexing. The principles have an ancient history going back at least to the thirteenth century when Bible concordances were first produced. Concordances are alphabetical arrangements of words occurring in a particular text or collection of texts like the Bible or the works of Shakespeare. Probably the one best known is Alexander Cruden's *Complete Concordance to the Holy Scripture,* first published in 1737 and still in print. Concordances can now be produced by computer programs which suppress designated words in the title and select significant words for indexing. Among the suppressed words are prepositions, conjunctions, and other words which do not convey significant information, such as "introduction" and "essay." These words make up what is called a stop list. Indexes employing this technique usually consist of a register of articles arranged in some sequenced order which can be readily accessed by the number assigned to the index term. KWIC indexing by machine was first proposed in 1959 and was adopted almost immediately by *Biological Abstracts,* which before this had used a controlled vocabulary of "subjects rather than words" along with a systematic index to reflect the taxonomic aspects of the subjects included.

Biological Abstracts, which has been published since 1926, is a primary indexing–abstracting medium in the life sciences. Along with its companion publication now called *Biological Abstracts/RRM* which indexes "reports, reviews, and meetings," the system adds over 300,000 items a year to its database, which now includes over 4 million items. The items are drawn from approximately 9,000 journals reviewed in the system, out of which all the articles in 3,300 are selected for indexing. *Biological Abstracts* appears in two issues a month which make up two volumes a year, each with a cumulated index of the keyword index and other indexes.

There are two major shortcomings associated with keyword indexes. One is that they tend to scatter references to the same subject, because of the lack of uniform terms. Thus, to scan all the titles with terms related to blood one would also have to look up ABO blood groups, Hematology, Hemolysis, Plasma, RH factor, Serum, and many other terms. The other shortcoming is that authors' titles do not always contain the necessary significant terms to define the content of the article. *Biological Abstracts* has attempted to overcome these difficulties in the first place by adding supplementary indexes and in the second place by editorially adding significant words to the author's title which are then permuted in the *Subject Index* along with the title words. An average of six terms is added to each title. The process is described as follows:

> This string of original title words and added keywords is then sorted alphabetically on each substantive term. Each significant term is printed in the index

in context; it is preceded and followed by that portion of the adjacent title word which will fit in the index column (i.e. 60 positions, including spaces). (198)

The process is perhaps best illustrated by taking an article and tracing it through its various permutations in the subject index:

Taylor D, Hockstein P. Reduction of metmyoglobin in myocytes. *J Mol Cell Cardiol* 1982, 14:133–40.

It appears in the following entries in their alphabetical order in the subject index:

T VENTRICLE MENADIONE	ADRIAMYCIN	DICOUMAROL METABOLIC-D
ROL METABOLIC-DRUG DT	DIAPHORASE	MET HEMO GLOBIN OXYGEN
MENADIONE ADRIAMYCIN	DICOUMAROL	METABOLIC-DRUG DT DIAP
MAROL METABOLIC-DRUG	DT	DIAPHASE MET HEMO GLOBIN OXY
REDUCTION OF MET MYO	GLOBIN	IN MYOCYTES RAT HEART VENTRI
DIAPHORASE MET HEMO	GLOBIN	OXYGEN BINDING/REDUCTION
LOBIN IN MYOCETES RAT	HEART	VENTRICLE MENADIONE ADRIAMY
UG DT DIAPHASE MET	HEMO	GLOBIN
S RAT HEART VENTRICLE	MENADIONE	ADRIAMYCIN DICOUMAROL M
DRIAMYCIN DICOUMAROL	METABOLIC-DRUG	DT DIAPHORASE MET H
ING/REDUCTION OF MET	MYO	GLOBIN IN MYOCYTES RAT HEART
RASE MET HEMO GLOBIN	OXYGEN	BINDING/REDUCTION OF MET M
YO GLOBIN IN MYOCYTES	RAT	HEART VENTRICLE MENADIONE ADRI
LOBIN OXYGEN BINDING	REDUCTION	OF MET MYO GLOBIN IN MYO
IN MYOCYTES RAT HEART	VENTRICLE	MENADIONE ADRIAMYCIN DIC

From these entries we can reconstruct the word string of original title words and added terms which was processed by the computer program:

REDUCTION OF MET MYO GLOBIN IN MYOCYTES RAT HEART VENTRICLE MENADIONE ADRIAMYCIN DICOUMAROL METABOLIC-DRUG DT DIAPHORASE MET HEMO GLOBIN OXYGEN BINDING

Counting backward and forward from the keyword for the appropriate numbers of characters and spaces will reproduce the subject entry.

The scatter of references in the subject index is compensated for to some extent by additional indexes which provide other access points which can also be coordinated with the entries in the subject index. The *Biosystematic Index* is a classified index arranged by broad taxonomic categories under which are listed the respective phyla, classes, order, etc. Concept terms such as "genetics and cytogenetics—human", "toxicology—pharmacological" are added to indicate the subject emphasis of the article. The taxonomic approach is also provided with an alphabetical list of the genus and species names covered in the titles or in the abstracts. Each of the terms in this Generic Index is also accompanied by a designation of the major concept in the article. Finally, in addition to an *Author Index* there is a *Concept Index* in which the relevant abstract numbers are posted under a broad subject which has been used to characterize the major emphasis of the article. Coordination of these indexes may offer some difficulties in the printed version, but as we shall see these are minimized in searching *Biological Abstracts* by computer.

Coordination

Almost every literature search is accomplished by means of some form of coordination. When you use a library card catalog, you coordinate visually the subject term or other names you have used to enter the catalog with the other words and information on the card to determine whether the item represented by the card may be useful to you. In using a KWIC index, you search by finding the most significant term which defines your query and by then coordinating it visually with the other terms with which it is displayed in context. In searching controlled vocabulary indexes, it is seldom that a single term will define precisely the subject you are investigating. Other terms are coordinated with it either by visual inspection of other words in the title in the printed index or programs in the computer which can perform this task automatically. This process is called "post-coordination."

The idea of term or concept coordination also has a long history. The French criminologist Alphonse Bertillon developed a system using this principle in the last quarter of the nineteenth century. He was assigned as a young man to the first section of the Paris Prefecture of Police, well known to readers of detective novels as the "Sûreté." His task was to file cards on known criminals and retrieve them on demand. Using information on human measurements he acquired from his father, a well-known anthropologist, he devised a system in which cards could be filed in a systematic order according to such factors as length and breadth of head and length of fingers. This system enabled him rapidly to identify criminals who had previously been entered into the file, by checking the coordinates of their measurements (436). A patent for a similar system was issued to a Mr. Taylor in 1915. As a birdwatcher he found the current handbooks inadequate to identify birds before they disappeared from view. He placed bird names on one series of cards and characteristics such as plumage, crest, and color on another series. When the appropriate cards were superimposed, the correct bird was identified by the light which showed through the holes punched in the card (233:44). These ideas were adapted to information retrieval systems around 1960 by what was called a "uniterm" system of indexing. A document or journal article number was posted on as many single-term cards as necessary to characterize its contents. If, for example, you wished to retrieve all the relevant citations in the file on *Kidney neoplasms* you pulled the card on *Kidney* and the card on *Neoplasms*. The matching numbers which had been posted on both cards would represent all the items concerned with both *Kidneys* and *Neoplasms.* One problem with the system was that the coincidence did not always represent the relationship between terms that was being sought. Before the computer began to play an important role in information retrieval, the principle was applied in various devices which were called "optical coincidence" systems or, with what seems like a sense of playfulness or even a suggestion of impropriety, "peek-a-boo" systems.

To reduce the number of matches or coordinations that needed to be made, some indexing languages precoordinate some of their terms, that is, use such terms as *Kidney neoplasms* as single terms. As with natural language and controlled indexing systems, there do not seem to be any pure examples of either pre-coordinated or post-coordinated systems. Pre-coordination is used even in natural language systems, as we have seen in *Biological Abstracts* and in other indexes such as the *Permuterm Subject Index* which is supplied along with the *Science Citation Index,* a novel form of index which I shall discuss in the next chapter. The *Permuterm Subject Index* is derived from the *source* articles which are published in each issue of the *Science Citation Index,* that is, the articles published in the current year rather than the articles which are *cited* in them. All the significant words in all the articles are listed alphabetically. Under each of these terms there is another alphabetical list of all the other significant terms in the articles in which the primary term is included along with the name of the author of the article in which those terms are included in the title. A search is conducted by coordinating the primary term with all the secondary terms in the list which are relevant to the inquiry and are identified by the name of the same author. Not all terms are included in the list; as with the KWIC index, there is a "stop-list" of suppressed words like "the" and "and." In addition, there is a "semi-stop" list of words which are useful as secondary terms to help define and narrow the search, but do not serve any purpose as primary terms, terms such as "method," "analysis," "area." All other terms appear as both primary and secondary terms. There are some other vocabulary conventions; British spellings are reconciled with American, and all foreign articles are indexed with cognate English terms. The principles can be illustrated by extracting the secondary terms under the primary term *Brain-tissue* from the *Permuterm Subject Index* for the following source article of a recent issue of the *Science Citation Index:*

Wightman RM, Bright CE, Caviness JM. Direct measurement of the effect of potassium, calcium, veratridine, and amphetamine in the rate of release of dopamine from super-fused brain-tissue. Life Sci 1981, 28:1279–86.

Under the term *Brain-tissue* in the *Permuted Subject Index* there is an alphabetical list of about 100 title words in alphabetical order, each identified by the name of an author in whose title the term appears. The terms which are identified by the name Wightman are the following:

Amphetamine
Calcium
* Direct
Dopamine
* Effect

    * Measurement
    Potassium
    * Rate
    * Release
    Super-fused
    Veratridine

The terms marked here with an asterisk are words in the "semi-stop" list, words which will not also appear as primary terms. All the others can be searched and will have the same array of terms including brain-tissue, which becomes a secondary term, as well as all the title terms for other source articles which contain the new primary term.

Controlled Vocabularies

Natural language indexes may reflect more precisely the language used in the literature, and may be produced more quickly and economically, than controlled language indexes, but they also provide problems in the searching process. In one study, a search on the zoological name for sheep, *Ovis,* as a title retrieved only 112 articles. Since this system permitted searches on title terms as well as controlled vocabulary terms, a search on the title term *Sheep* produced 2,991 articles. However, when the system was searched with the term *Sheep* as a controlled vocabulary term, 4,507 articles were retrieved (331). Another study surveyed the literature on *Apanteles,* a natural enemy of the Asiatic rice worm borer, by using keyword indexes based on title words and words in abstracts and found that the species name was mentioned in only 3 percent of the titles of relevant articles and in only 20 percent of the abstracts. The investigators also found numerous variations in the spellings of the genus and species name (402).

Controlled vocabularies are designed to overcome these problems by designating preferred terms for concepts, and by referring to these terms from synonyms, spelling variants, and narrower and broader terms. Policies are also established to deal with such matters as abbreviations, acronyms, and the form of the subject heading. Lancaster has listed these criteria for the effectiveness of controlled vocabularies:

1. They should be based on the language in the literature and on the needs of users.
2. The terms should be sufficiently specific so that irrelevant items are not retrieved.
3. There should be enough compound (pre-coordinated) terms to prevent false coordination.
4. They should promote consistency in indexing and searching by controlling synonyms, near synonyms, and quasi-synonyms.

5. They should reduce ambiguity through distinguishing between homographs and by defining ambiguous terms.
6. They should assist the user in selecting appropriate terms through classification and a cross-reference system (271).

These criteria are exemplified by the indexing language that is used by MEDLARS (the Medical Literature Analysis and Retrieval System) of the National Library of Medicine, which produces the printed *Index Medicus*, its online version MEDLINE, and a number of other products which provide an important resource for the life sciences. The computer-based system called MED-LARS on which these products are largely based was inaugurated in 1960, but it was preceded by a number of other indexes produced by the Library going back to 1879, which are covered in some of the sources listed in the Appendix. *Index Medicus* is published monthly and cumulated annually in ten or more volumes under the title *Cumulated Index Medicus*. At the end of fiscal year 1981, *Index Medicus* was covering 2,664 journals in many languages covering all areas of clinical and experimental medicine, as well as the associated basic sciences, dentistry, nursing, public health, and the allied health sciences. These journals are selected from the 25,000 currently received by the Library with the help of a panel of consultants who meet periodically with the library staff to review new titles for inclusion. There was a brief period in which selected multi-authored monographs were included, but this was discontinued in order to be able to accommodate the increasing number of high-quality journals which were appearing. In fiscal 1981, 256,112 items were added to *Index Medicus*, of which almost half included abstracts in MEDLINE (324). In 1981 it was estimated that an average of 69 days elapsed between the time "a top priority journal" was received by the Library and the time it was entered into MEDLINE (336:113). The elapsed time was obviously higher for lower priority journals and the printed version, which may be 6 months or longer.

MEDLARS uses a highly controlled language called *MeSH* (Medical Subject Headings), a list of preferred or accepted terms. Lists of this kind may also be referred to as a subject authority list or thesaurus. A thesaurus may include a cross-reference system to refer from synonymous terms which are not used to those that are, and to broader and narrower terms under which concepts may be indexed. Terms, or "descriptors," may consist of single words, or two or more words which are combined, hyphenated, inverted, or subdivided by sub-headings, depending on the conventions of the indexing language. MeSH uses most of these variants. The expanded indexing vocabulary available in MED-LARS consists of some 25,000 terms, of which only 10,000 are used in *Index Medicus*. The other terms consist of check tags which refer to age groups, experimental animals, and other aspects of an article that are available only in computer-based online searching. They also include so-called minor descriptors which also can only be searched online. The *Index Medicus* MeSH vocabulary appears in a revised edition at the beginning of each year, incorporating all the changes which are being implemented in that year. It also includes categorized or classified lists of terms called "tree structures" which provide additional approaches to the vocabulary and are useful for online searching.

Other versions of MeSH which include the full vocabulary are produced as specific aids to online users. *Medical Subject Headings Annotated List,* also published every year by the Library, includes notes for indexers and users which help to define terms as used in the system. A history of the use of the term is also included when required, which is useful when covering long periods of the database in searching.

The MeSH vocabulary is derived (i) from the literature, i.e., from such authoritative terminologies as *Bergey's Manual of Determinative Bacteriology* which are listed in the preface to each list and other authoritative sources, (ii) through the participation of professional associations, and (iii) from consultants. Terms appear as single words, *Fever;* homographs defined, *Aspiration (Psychology);* adjectival phrases, *Surgical flaps;* adjectival phrases inverted, *Surgery, plastic;* combined terms, *Missions and missionaries;* and as prepositional phrases, *Medicine in art.* Controlled vocabularies such as MeSH tend to be pragmatic rather than logical. Although official terminologies are consulted, they are not followed without variation and common usage tends to prevail. For example, the official *International Anatomical Nomenclature* term is used for the *Corpus collosum* (great commissure of the brain), but not for the *Pineal body* (epiphysis cerebris). Some of the choices may seem capricious, for example, *Addison's disease* (melasma suprarenale) but not Addison's anemia (see *Anemia, pernicious*). The term *Leprosy* is used instead of Hansen's disease, but this may be because the proprietary interest in the disease is still being disputed between the followers of Hansen and Neisser. These may all, however, be said to reflect common usage.

Like other controlled vocabularies MeSH incorporates a cross-reference system to refer from terms not used to terms which are used in their place (Hansen's disease, see *Leprosy*), from narrower terms which are not used because the volume of literature does not justify their use in the system (Barley, see under *Grain*), and to related terms (*Ethics, medical,* see also related *Human experimentation*). The controlled language of the National Agricultural Library, the *Agricultural-Biological Vocabulary,* also refers from narrower (specific) terms which are used to the related broader terms, but in MeSH these relationships appear only in the categorized list. Consulting the categorized list, however, supplies you with other related terms as well. MeSH also refers to abbreviations which are used from the full term which is not used (Deoxyribonucleic acid see *DNA*), or vice versa (DDVP see *Dichlorvos*). It refers to spelling variations (*Hemophilus* see *Haemophilus*), to semantically related terms (*Cardiology,* used for literature on the specialty, see *Heart disease* used for clinical aspects), from inverted forms to non-inverted forms (*Inhibitions, neural* see *Neural inhibitions*) and also the other way (*Fluorescence microscopy,* see *Microscopy, fluorescence*).

Another method of subdividing lists of references under a specific term in the printed *Index Medicus* is by means of subheadings which are used with various subject headings to further define the content of the article. They play a different role in online systems where they serve as additional search coordinates. There are seventy-eight subheadings in the MeSH vocabulary from *Abnormalities* (used with organs for congenital defects) to *Veterinary* (used for naturally occurring diseases in animals). The subheadings can only be used with designated categories. For instance, the subheading *Congenital* can be used with all the diseases in category C except C16, *Diseases, neonatal,* for obvious reasons. Subheadings must, however, be used with caution, because some permitted combinations like *Fatigue,* a C23 term (i.e., Fatigue—congenital) may seem like a judgmental comment rather than a recognized disease entity. *Physicians— adverse effects* is not a permitted use of the subheading (used only with categories D, E, F4, G3, H, and J), although the vocabulary does supply the term *Iatrogenic diseases* which may essentially mean the same thing.

Thelma Charen, one of the principal architects of the MeSH vocabulary, describes MEDLARS as a "system of coordinate indexing."

> In coordinate indexing the concepts in the texts of articles are expressed by the combination or coordination of two or more indexing terms, some destined for publication in *Index Medicus* and others destined for storage within the computer for future retrieval. (73)

MEDLARS, however, is not entirely a post-coordinated system. Many concepts in the vocabulary are represented by pre-coordinated terms. For instance, you would not look under *Pregnancy* (G8.520.769, a subclass under G8, *Physiology— Sex, Reproduction, Urogenital*) or *Diabetes* (C18.452.297, a subclass under *Nutritional and metabolic diseases*) for literature on diabetes in pregnancy, but under the pre-coordinated term *Pregnancy in diabetes* (C13.703.766, a subclass under *Pregnancy complications,* another pre-coordinated term). The association between *Liver cirrhosis* and *Alcoholism* is not sought by coordinating these two terms but by searching under *Liver diseases, alcoholic.* There are also the pre-coordinated natural language terms such as *Albuminuria* and *Proteinuria* (the presence of albumin or protein in the urine) which appear in the MeSH vocabulary.

There are other conventions which must be learned as well, as is true with all indexing languages. General articles on the diseases of an organ, for instance, are indexed under the name of the organ in MeSH *(Eye diseases, Heart diseases),* while articles on the respective specialties are indexed under the name of the specialty *(Ophthalmology, Cardiology).* Diseases caused by organisms are indexed as precoordinated headings under the disease, *Hookworm infection,* instead of under the organism, Uncinaria stenocephala (see *Hookworms*), unless the organism is specifically a subject of the article (331).

Controlled vocabularies, like natural language vocabularies, do not remain static. Terms for new concepts are constantly being introduced, and concepts formerly indexed under broad headings may be assigned to narrower headings. Users must be aware of these changes as they search the index from one year to the next. The MeSH vocabulary is revised every year and published with the first issue of *Index Medicus* for that year. New headings introduced for the first time are listed along with the terms under which these concepts were previously indexed. A cumulated list covering all the changes since 1966 was published in 1971 and is supplemented every year. Each list indicates the span of years for which the concept was covered under a deleted or alternative term. These changes are also noted in the main MeSH vocabulary. The 1983 MeSH provides examples of many of the kinds of changes that are made, from specific terms which are no longer justified (*Ejection seats,* last used in 1978) to related general headings (*Aircraft,* under which articles on *Ejection seats* if they occur in the literature must now be searched). On the other hand, when justified by the volume of literature, concepts which were indexed under more general headings may be assigned specific headings (*Antibiotics, lactam* previously indexed under *Antibiotics*). *Mice, nude* and *Mice, obese* as more specific terms arrived in 1975, but we had to wait until 1978 for *Mice, quaking.*

Sometimes terms are changed to reflect a new point of view as when *Sterility* was replaced by *Infertility.* Terms also are introduced which have no clearly identified predecessors. *Behavioral medicine* defined in the *Annotated MeSH* as "The interdisciplinary field concerned with the development and integration of behavioral and biomedical science knowledge and techniques relevant to health and illness . . ." was introduced in 1983. Before that, articles on this emergent medical specialty had to be sought out under a variety of headings. Changes of this kind occur with organisms and chemical compounds for a variety of reasons. *2-Acetylaminofluorene* was introduced in 1983 to replace *Acetylaminofluorene,* under which articles on the subject appeared from 1980 to 1983. Before, it had overlapping coverage under the term *Fluorene* from 1964 to 1979 and also under *Acetamides* from 1973 to 1974. Changes also occur with check tags and minor descriptors, but they do not affect searches in the printed *Index Medicus* except when minor descriptors become major descriptors. They do, however, affect searching on MEDLINE. Computer searching, as we shall see in the next chapter, introduced many new capabilities. In fact, some of the distinctions between controlled and natural language vocabularies have become less meaningful.

Other Vocabularies

Most indexing languages differ considerably from one another whether they are controlled or uncontrolled, post- or pre-coordinated. The vocabulary of *Agricola* (Agricultural On Line Access) of the U.S. Department of Agriculture, which has also produced the *Bibliography of Agriculture* since 1942, is a controlled vocabulary system similar to that of MEDLARS. However, it uses different conventions, and the terminology also shows many variations. The database is highly oriented toward agriculture in its applied and scientific aspects, but also contains information of interest and value to investigators in the life sciences. *Agricola* covers over 1,200 journals and adds about 10,000 citations a month. The thesaurus uses the same kind of term relationships as MeSH, VF (*used for, see in* MeSH), NT (*narrower term, see under* in MeSH), and RT (*related term, see related* in MeSH), but *Agricola* also uses BT (*broader term*) to refer from all related narrower terms, while the categorized or classified list in MeSH must be consulted for this information. *Agricola* does not use inverted headings, with the result that some related terms are scattered, *Fad diets* (*Diet fads* in MeSH) and *High altitude diets;* MeSH brings such terms together by inversions, *Diet, reducing* and *Diet, sodium restricted.* There are many terms in *Agricola* which are not used in MeSH and vice versa. In short, they are languages which must be learned independently.

When it comes to indexing languages such as that used in *Chemical Abstracts,* it is difficult to characterize them as either controlled or uncontrolled. Searching the chemical literature, in any case, is a complex and intricate task which must be mastered. However, since chemistry pervades almost all the areas of the life sciences, *Chemical Abstracts* provides one of the principal searching sources for the life sciences literature. It has been published by the American Chemical Society since 1907 and has grown and been modified considerably since that time. The growth of the chemical literature is exemplified by the fact that it took 30 years to produce the first million abstracts; now it takes about 2 years to add the same number (20). Each of the weekly issues covers about half of the eighty subject sections into which the abstracts are classified. Over 14,000 journals, as well as conferences, proceedings, and patents, are screened for chemical articles, producing about 500,000 abstracts a year published in two parts, each of which encompasses many volumes.

An elaborate indexing structure has been created which has gone through several modifications since it was inaugurated. The weekly issues have three indexes: keyword, author, and patent. The weekly keyword index uses an uncontrolled vocabulary derived from one or more significant words in the title. All synonymous forms must be searched. Terms may be added at the discretion of the indexer but in no standard form. The article may be entered under a single concept or under several concepts. These weekly indexes bear no relationship to the subject indexes in the semiannual volumes, which are based on a kind of controlled vocabulary.

Comprehensive indexes, which appear about 6 months after the completion of each of the semiannual volumes, completely supplant the weekly indexes. They include (i) a Subject Index which includes all terms except specific chemical substances (classes of substances, *Carbonates,* for example, are included, but not the specific substance, *Carbon*); (ii) a Chemical Substance Index which uses the standard systematic nomenclature; (iii) a Formula Index based on the order of the molecular formula; (iv) an Index of Ring Systems which provides a means of determining the systematic name when the ring system of the substance is known; (v) an Author Index; and (vi) a Patent Index and concordance which is cumulated from the weekly issues. An additional approach to the difficult nomenclature of chemical substances is provided through unique registry numbers which have been included in abstracts and indexes for all chemical substances since 1965, and which links the different names used for various chemicals. By consulting the *Registry Handbook* with a given number, one can determine the systematic name of a substance and its molecular formula.

Collective indexes to *Chemical Abstracts* have been published, at first for each 10-year period beginning 1907 to 1916, and then for 5-year periods starting with 1957 to 1961. The *Index Guide,* which is the key to the controlled vocabulary, was formerly published at the beginning of each 5-year period, with cumulative supplements covering the intervening years published once a year. Currently, the *Index Guide* is being totally revised three times during the 5-year period. The index lists all the preferred terms along with references from synonyms and other terms.

It is difficult to try to reverse the indexing process by tracing a particular abstract through any of the indexes, but it may be informative to list those which were found for the following example:

Milner RDG, De Gasparo M. Effect of nonessential amino acids on fetal rat pancreatic growth and insulin secretion in rats. *J Endocrinol* 1981, 91:289–97.

The article was listed under the following entries in the weekly keyword index:

Amino
    acid nonessential embryo pancreas

Insulin
    embryo nonessential amino acid

Pancreas
    embryo nonessential amino acid

There were no entries under *fetus* or *fetal* or *rat* because these apparently were nonessential aspects of the article. The entries in the semiannual *Subject Index* were:

Amino acid, biological studies
    nonessential
        insulin secretion by pancreas of embryo in culture in response to

Pancreatic islet of Langerhans
    insulin secretion by
        of embryo in culture, nonessential amino acid effect on

There was no entry in the *Subject Index* for insulin because it is the name of a specific substance. The *Chemical Substance Index,* however, had the following entry for the article:

Insulin, biological studies
    release of
        by pancreas of embryo in culture, nonessential amino acid effect on

The indexing language can, therefore, be described as one which uses preferred terms only for the primary terms, which are followed by natural language terms which provide a kind of mini-abstract of the article.

Another vocabulary which is difficult to characterize is that of *Excerpta Medica,* the largest English language abstracting service devoted to medicine and the related disciplines. It was inaugurated in 1946 by a group of Dutch physicians and acquired by the Elsevier Publishing Company in 1972. More than 4,400 primary journals are regularly screened for relevant articles. The citations are divided into forty-three abstract journals, each covering a particular medical specialty or biomedical discipline. About 250,000 records a year are added to the system, but only 150,000 of them appear in the printed abstract journals. The others can be searched in an online version of the database. About 60 percent of the citations include abstracts, and on the average each record appears in two different abstract journals. The average lag from the time a journal is published until the articles appear in one of the abstract journals is said to be about 9 months, which is reduced to 2 months for the online version. The system also produces two indexes in addition to the abstract journals, the *Drug Literature Index* and *Adverse Reactions Titles.* You can determine which abstract journal to consult by checking the *Guide to Excerpta Medica Classification and Indexing System,* which will reveal, for example, that articles on *Color vision* are to be found in Section 2, *Physiology* (subsection 14.1 *Visual receptors*) and in Section 12, *Ophthalmology* (subsection 30.2 *Visual acuity and vision*).

Each issue of the abstract journals has its own subject and author index, which is cumulated in the last issue for that section. There are, unfortunately, no collective annual indexes for the entire series, although this difficulty is surmounted to some extent by the general subject index to the sections in the *Guide.* Although the *Excerpta Medica* vocabulary, MALIMET (master list of medical indexing terms), bears some superficial resemblance to MeSH, it is a separate and distinct indexing language. It is said to contain 200,000 controlled or preferred terms and over 250,000 synonyms. These seem rather large numbers when one considers that *Webster's New International Dictionary* contains in all about 550,000 entries. The mystery is solved when we discover that all compound terms are counted as separate terms and find, for example, that there are about 100 terms which begin with the word "abdominal," from *Abdominal abcess* to *Abdominal X-ray.* Synonyms include variant or inverted forms of terms as well as true synonyms; for example, the synonyms listed under the preferred term *Brain tumor* are *Brain tumour, Cerebral tumour, Cerebral tumor,* and *Tumor, brain.* Each index entry consists of primary terms which are permuted in their alphabetical place in the index and secondary terms which may not be part of the preferred vocabulary but add information about such things as dosage level and number of patients studied. Tag codes which are similar to the MeSH subheadings and check tags are also used to define such things as age group, experimental animal, and routes of drug administration. These also are searchable only in the online version of *Excerpta Medica.*

Although the *Excerpta Medica* vocabulary contains many terms which are similar to those in MeSH, it uses different conventions and tends toward more popular terms, *Slaughterhouses,* for example, instead of the MeSH *Abbatoirs.* Terms are not inverted as in MeSH, for example, *Acute abdomen* rather than the MeSH *Abdomen, acute.* A typical index entry is that for the article:

> Glass ML, Buggren WW, Johansen K. Pulmonary diffusing capacity of the bullfrog. *Acta Physiol Scand* 1981, 113/4:485–90.

An index entry for this article is:

> Carbon monoxide, lung diffusion capacity, lung perfusion, lung volume, oxygen consumption, mass spectrometry, frog.

Index entries will be found for all the other terms in the string with the exception of *mass spectrometry* and *frog.* Mass spectrometry is a preferred term but here apparently is not regarded as deserving an index entry, and the term frog is a tag which is not indexed in the printed version.

Classification: Generality and Specificity

Another method of organizing and controlling literature which also has an ancient lineage is classification. The process of arranging ideas and objects in classes, whether they are books or organisms, is, of course, basic to all intellectual activity. This is particularly true in the so-called taxonomic sciences such as bacteriology, botany, or zoology, but also in such areas as nosology or disease nomenclature. "Classifying things is perhaps the most fundamental and characteristic activity of the human mind, and underlies all forms of science," says a writer on biology and classification. One of the distinguishing characteristics of science is its ability to quantify and measure. He says:

> Neither counting or measuring can, however be the most fundamental processes in our study of the material universe—before you can do either to any purpose, you must first select what you propose to count or measure, which presupposes a classification. (98:2)

Words are in themselves, as semanticists remind us, frequently designations of a class, unless they are used to designate a specific organism or object—that specific individual called John Smith or that specific book lying there on the table. This factor may lead to semantic ambiguity, prejudice, and confusion, but it also enables us to cope with the material world around us.

Classifications play a major role in indexing languages. First, in controlled vocabulary systems it enables us to index concepts at different levels of specificity or generality. Even in natural language systems one must be aware of the specific (species) relationships of words to broader concepts. These relationships are taken into consideration in indexing thesauri by displaying terms in both alphabetical and classified order, by referring from broad terms to narrow terms and vice versa, and by supplementary approaches such as term inversion, which are in effect hidden classifications.

Logicians suggest rather strict guidelines for classifications:

1. The basis of the division must be *clear* and unambiguous at every stage.
2. The division must be *exhaustive* at every stage.
3. The division must be *exclusive* at every stage.
4. Whenever possible the component parts of each division should be of *comparable* rank.
5. At each division there should be a single and *uniform* basis of division.
6. The division should be justified at each step in terms of a *single governing purpose*.
7. The classification should be *informative* (381).

These criteria are difficult to apply even in dealing with highly discrete phenomena which are easily recognized. Most classifications used in information systems, such as the familiar Dewey Decimal or Library of Congress classification used for arranging books by subject in libraries, are highly pragmatic and functional. They are designed to serve a specific purpose and embody numerous compromises. Another characteristic of most classification systems is that they are unilinear or undimensional, whereas all objects and ideas exist in a multi-dimensional world. The act of classification involves the selection of one characteristic out of an array of characteristics, in order to place it in relation with other objects or ideas based on the same characteristic. Tomatoes, for example, are classified as a vegetable in cookbooks and as a fruit in botanical books. A good example in a library collection would be a book which describes a statistical technique using a form of word analysis to determine whether Thomas à Kempis might be the author of *The Imitation of Christ*. The library which acquired the book for a member of its theology faculty might classify it with its books on religion, but if the emphasis was on the statistical methods employed it could easily be placed with books on mathematics.

Subject headings and other terms in indexing languages are explicitly or implicitly parts of a hierarchical structure in that most of them can be subsumed or subordinated to other terms. In controlled vocabulary systems this is exploited by using inverted headings or adjectival phrases. For example, if you had articles on *Big boys, Small boys, Big girls,* and *Small girls* you could bring the boys or girls together by inverting the headings. On the other hand, by not inverting the terms you could leave the *Big boys* with the *Big girls,* which under some circumstances might not be entirely inappropriate. MeSH does this constantly, for example, by bringing *Neoplasms* together with such inverted headings as *Neoplasms, Muscle tissue* and *Neoplasms, Nervous tissue,* or *Neurosis, Anxiety,* and *Neurosis, Post-traumatic.* MeSH does not do this consistently since it also contains such terms as *Breast neoplasms* and *Kidney neoplasms.* The MALIMET thesaurus of *Excerpta Medica,* as we have seen, brings together over 100 terms relating to the digestive system by using the combining adjective *Alimentary.*

The levels of specificity or generality of terms used in the controlled vocabularies depend on a number of factors. One is the size of the entire collection, either as a totality or in terms of the size of the particular subcollection. For example, if there are only 100 items in a collection, ten descriptors (classes) would divide it into ten groups, each of which could be scanned easily for specific descriptors and yet bring related items together. However, if the collection grows to 1,000 or 10,000 units, ten classes would not divide it sufficiently for efficient searching. Although indexers for *Index Medicus* are instructed to index at the greatest level of specificity consistent with the system, the *Index* could easily become unmanageable if, for instance, a heading was used for each organism which is cited in the literature. The class Mammalia alone contains 4,500 species, Mollusca about 100,000, and Insecta 700,000. Terms are, therefore, introduced into indexing vocabularies in accordance with the frequency with which they are needed to characterize the content of the articles indexed (331). This will vary from index to index according to the nature of the literature and the needs of a specific clientele. The *Agricultural Index,* for example, has entries for *Cocker Spaniel dogs, Collie dogs, Schnauzer dogs,* etc., whereas MeSH recognizes only *Dogs.* The existence of a general heading does not imply that related specific headings are not also used. The MeSH term *Artiodactyla* is used for some animals that belong to the order such as *Bison, Giraffe,* and *Hippopotamus* which are not even granted status as cross references; other animals in the same order which are also indexed under the general heading, *Bears, Lions, Otters,* and even *Skunks,* do achieve this distinction. *Camels, Deer,* and *Goats* as well as others which belong to the same order are afforded separate status in the vocabulary (331). These terms are all brought together in one display in the categorized list under B2, Organisms, Vertebrates.

The MeSH categorized list is now published with the designation "tree structures" to characterize the branching nature of classifications. All terms used in the vocabulary are assigned to one or more of fifteen general classes from A (Anatomical structures), to B (Organisms), to Z (Geographic names which are not searchable in *Index Medicus* but can be used as coordinates in searching MEDLINE). Each term is assigned a numerical code to designate its place in each subclass. Spiders, for instance, have the code designation B1.131.166.803 representing their position in the subclass 166, *Arachnida,* of the subclass 131, *Arthropoda,* under B1, *Invertebrates.* Each of these terms can be searched separately or collectively in the computer-based system, as we shall see later. These relationships are not quite as clear in some of the other classes such as C, *Disease,* or D, *Chemicals,* in which some subclasses may have as many as three alternative locations.

The *Excerpta Medica* system uses classification in quite a different way. The abstracts in each issue of the abstract journals are arranged in a classified order to make browsing and searching easier. Each of these subject classes is assigned numerical codes to designate over 5,000 categories which can be used for searching. As in MeSH these classes can be assigned more than one position in the classification. *Chemical Abstracts* and *Biological Abstracts* also issue their printed abstracts in a classified order. *Chemical Abstracts* includes thirty-four sections covering biochemistry and organic chemistry one week and the remaining forty-six sections in the alternate week. In addition, the *Index Guide* supplies an arrangement of the general subject headings in a quasi-hierarchical order which codifies the relationships between the terms used in the controlled vocabulary. They are displayed in fifty-eight groups in alphabetical order from *Agriculture* to *Universe*. *Rocks*, for example, find their place at the seventh level of specificity under Universe, Stars, Solar system, Planets, Earth, and Geological deposits. *Biological Abstracts*, as we have seen, uses its *Biosystematic Index* to bring species terms together and its *Concept Index* to relate the other terms. It is frequently necessary to use these alternative approaches, because as one investigator discovered, relevant papers can be scattered through various sections of *Biological Abstracts*. He searched a 5-month period of the abstract journal and located 2,300 virology-related papers in twenty different sections (71).

These hierarchical arrangements and classifications become vastly more important in computer-based and online searches, which can be performed on all the terms in a particular category. For example, if you were interested in finding information in MEDLARS on all forms of *Skin manifestations* (C17.871) as *adverse effects* (subheading) of all *Antibiotics* (D20.85), you might wish to coordinate all the *Skin diseases* terms, well over 100, with all the *Antibiotics* terms, close to 200. It would be extremely tedious and time consuming to search *Index Medicus* manually with this strategy. The task becomes relatively easy with the assistance of the computer, as we shall see in Chapter 10, "Searching the Literature." In fact, the arrival of the computer changes the entire environment of literature searching and tends to break down to some extent the distinctions which are made between indexing languages.

Chapter 9

Citation Indexing and Analysis

There is a method of indexing the literature which I did not discuss in the chapter on indexing language because, unlike the others, it is not dependent on words. In fact, one of its major strengths is that it avoids all the ambiguities and complications inherent in word indexing. This method is based on the use of the references or citations which almost every scientific author uses to give credit to predecessors or to support and document statements made. The terms "reference" and "citation" are sometimes used interchangeably, but it is useful to distinguish between them in order to understand more clearly the process of citation indexing. "A reference," says one author, "is the acknowledgement that one document gives to another; a citation is the acknowledgement that one document receives from another" (414:106). In other words, it can be called a reference when it is in its place in the article calling attention to earlier work and a citation when it is separated from the article and can be used to call attention to a later work in which it was used as a reference. This is, in fact, one of the unique contributions of the technique called citation indexing. Having a paper or a citation to a paper in hand, by means of citation indexing you can discover whether it has been cited by subsequent authors. Thus, you can have access to later work which is theoretically on the same or a related subject.

Citation indexing, therefore, is based on the assumption that references in a paper are relevant to the topic the author is discussing and that if you can find others papers which cite the same reference they may also be relevant. Citations thus become equivalent to subject headings or the number of coordinated subject headings it may take to characterize a problem. "Citations," says Eugene Garfield, who developed citation indexing in science as a new form of access to the literature, "are the formal, explicit linkages between papers that have particular points in common" (150:1). Ziman echoes and expands the concept. "Citations," he says, "not only vouch for the authority and relevance of the statements they are called upon to support, they embed the whole work in a context of previous achievements and current aspirations" (475). Further assumptions implied in the use of citation indexing are (i) that the authors have referred to all the important documents and (ii) that they are all relevant (414:87), that is, that the authors have not sinned either by omission or commission.

These assumptions have all been questioned, as we shall see later. Citations, says Merton, operate within both a cognitive and a moral framework. As a sociologist he would certainly also add the social framework in which recognition is bestowed and rewards in science are allocated. Cognitively, he says, they provide the historical linkages in knowledge and guide readers to appropriate sources. Morally, they are payments of intellectual debts in the only form this can be accomplished, through the acknowledgment of sources (307:viii). It is also recognized that other factors may be involved: the familiarity of the author with the literature, language preferences, personal loyalties, and even self-interest. Further, a reference may be judged to be useful or relevant to the author but may not prove so to the inquirer.

In spite of these reservations, the merit and value of citation indexing has been recognized by the scientific community. This has been demonstrated by its widespread use in the past 20 years since it was first introduced as a means of information retrieval in science. There is good reason for this. Vocabulary-based indexes depend on the ability of the indexing languages to identify subject content. In controlled vocabularies this is a reflection of the effectiveness of the indexing language and the ability of indexers to use it. In natural language indexes it depends on the author's ability to describe the content of an article through the choice of title terms. Citation indexing, furthermore, is "semantically more stable than conventional subject indexing" (150:41), because it indicates intellectual relationships which are not based on words. Words tend to be used differently in different indexing systems and by different authors. It is also based on a judgment of relevance made by a specialist, the author, rather than by an indexer. The citation can thus be considered a surrogate for an indexing term or any combination of indexing terms it may take to express the subject content of a document.

"All searching by key-words, etc. is retrospective whereas scientific discovery requires prospective insights," says Bottle (51:260). The statement can be interpreted in various ways. First, pragmatically, it means that conventional indexes look back at what has already been published, whereas citation indexes look ahead to what is published after the article represented by the citation has appeared. Another interpretation could be that, once a name has been assigned to a concept, the relationship between the two has been established as a part of a consensus, while the progress of science depends on the establishment of new relationships. The ability to look ahead from an article to an article which cites it can also show whether the ideas contained in the earlier article have "been confirmed, extended, improved, tried, or corrected" (150:41). In this sense citation indexes provide quality filters for the literature. This is reflected in the fact that only one-half of all the available papers are cited in any year and that a large majority (72 percent) of all the papers published are cited only once (368). Citation indexes can also be more timely since they do not depend on the need for trained indexers to use critical judgments in selecting terms to apply to articles before they can be entered into a system. This factor is also related to the costliness of the indexing process.

Citation indexing had a long history of use in law before it was applied in science. This intellectual debt has been fully acknowledged in the literature (150:7). *Shepard's Citations* was founded in 1873 by Frank Shepard to index legal cases in Illinois. It now covers all the legislative and judicial branches of the U.S. government and the states. The system is used, says Jacobstein, "to determine that any given case that is to be relied on as authority is indeed still good authority. The decision must be checked to make positive that it has not been reversed by a higher court, or overruled by a subsequent decision of the same court" (232:226). A citation is defined in law as a "reference to authority necessary to substantiate the validity of one's argument or position" (232:xvi). Lawyers preparing cases to present before the courts must cite authorities to support their arguments. After locating relevant court decisions, statutes, or other legal precedents, they must take the further step to determine that the precedents can still be relied upon as good authority, that is, that they have not been overruled, reversed, or changed in subsequent actions. This is accomplished by "Shepardizing," that is, looking into one of the Shepard indexes to see if the precedent has been cited. In the process other cases are also uncovered which are relevant to the inquiry (6). The role the *Science Citation Index* plays in science is analogous to that of *Shepard's Citations* but is greatly extended.

The *Science Citation Index*

The *Science Citation Index* was introduced in 1964. One author has referred to it as the scientist's *Debrett*, alluding to the guide to barons, privy councillors, and peers of the realm which has been published in England since 1769 (97:16). The implication is that its function is to establish a scientist's standing in the hierarchy of the profession. It may be used for this purpose, although, as we shall see later, its primary objective is to serve as a guide to the current literature. Like *Shepard's Citations* it can also be used to discover whether scientific findings have been updated or modified. It is published every 2 months to cover the literature which appeared in that period and appears in three sections, a *Source Index* covering the current articles, a *Citation Index* made up of the references which appear in these articles, and a *Permuterm Index,* a word index to the current articles which I discussed earlier. The bimonthly issues are all cumulated annually. In addition three 5-year cumulations have been published covering the years 1965 to 1969, 1970 to 1974, and 1975 to 1979. These supersede the annual cumulations and reduce considerably the number of volumes to be consulted, since an article published in 1962 could have been cited at any time during the 15-year period. The indexing technique has also been extended to disciplines outside the sciences through the *Social Sciences Citation Index* and the *Arts and Humanities Citation Index.* A retrospective *Science Citation Index* covering the scientific literature which appeared from 1955 through 1964 has also recently been published.

The *Science Citation Index* covers the scientific and technical literature in agriculture, biological and environmental sciences, engineering, and medical, physical, chemical, and behavioral sciences. It is therefore more effective in multi-disciplinary searches than indexes which cover single disciplines. The *Index* and the other services the system supports would have been impossible without the sophisticated use of computer technology. The three 5-year cumulations cover about 6 million source articles published between 1965 and 1979 and over 25 million citations derived from these articles. The citations cover a much larger time span since literature of any period may be cited. Source articles are currently being added to the database at the rate of over half a million a year which represent over eight and one-half million citations.

Coverage in the *Science Citation Index* has increased constantly, from 1964 when 610 journals were indexed to over 3,000 in 1981 in addition to 635 "non-journal" sources which included multi-authored books. Despite the fact that this number of journals represents less than 10 percent of the estimated world scientific journals, studies of scattering, such as those referred to in Chapter 5, indicate that 75 percent of all references occur in fewer than 1,000 journals and that 2,000 journals account for 84 percent (150:21). The *Citation Index* includes references to a great many more journals, since every reference is indexed with an indication of the source article in which it appeared. These include all references, whether they are to journal articles, editorials, letters, reports, theses, patents, or books. This may lead to interesting results, particularly in the case of articles published in scientific spoofs, such as J. S. Greenstein's article (193), which dealt with an alleged deodorant which had been accidentally discovered to have spermicidal effects. Among the references from this article which appeared in the *Citation Index* for 1965 were:

Yolk A, White B. On slicing a hard boiled egg. *Popular Mechanics,* 1948, 39:251.
Sturm A, Drang B. *Population control and human sexuality.* New York, Harder and Harder Press, 1961.
Gabriel A. Personal communication.

Although the *Science Citation Index* is multi-disciplinary, a recent study has indicated that the coverage of the life sciences literature is excellent. One investigator analyzed the references in seventy dissertations written between 1973 and 1977 in anatomy, biochemistry, immunology, microbiology, pathology, pharmacology, and physiology. He found that the coverage was fairly uniform in all these disciplines, with a range of 92.24 percent for pathology to 96.75 percent for microbiology (365). Another analysis in 1981 concluded that the *Science Citation Index* was "an excellent, internationally balanced data source for the core of the physical and biological sciences, particularly for the English speaking countries" (69).

The *Source Index* is arranged alphabetically by the name of the first author of all the articles in journals and other multi-authored works published in the period covered by the index. The source items include the names of secondary authors, journal title abbreviation, volume, page, and year of the article, the title in English, and the number of references it contains. Anonymous articles are listed before named authors, arranged alphabetically by journal abbreviations and in chronological order under each journal. There is also a "corporate index" which identifies where the work represented by a particular source paper was performed and can be used to determine the activities occurring at particular institutions.

The *Citation Index* includes all the references listed in the sources articles, no matter what year they were published. They are listed alphabetically by the article's first author. All the author's cited articles in the period covered are arranged chronologically under the author's name, with the year, volume, page, and journal title in which the article appeared. Under each of these cited articles are listed the source articles in which they appear as references. As there is an average of fourteen references in each source article (368), any one of the cited articles may have anywhere from a single to several hundred source articles listed under it.

The *Permuterm Index* which also appears with each separate issue or cumulation of the *Science Citation Index* can be used as a subject guide to the articles in the *Source Index,* but it can also be used as a means of entry into the *Citation Index,* by selecting one of the key references in a source article for that purpose.

One of the first places new users of the *Science Citation Index* frequently look is in the *Citation Index* to see if anyone has cited their work. This, however, can be more than an ego trip; it may, in fact, prove deflationary to find that your work is not cited at all. The *SCI Annual Guide* describes the process as one "used by scientists to determine whether their work has been applied or criticized by others" and adds that it "can also facilitate feedback in the communications process" (396). The process also illustrates the usual starting point for an inquiry into the *SCI:* an earlier citation on the subject being investigated or the names of authors who have been identified with the subject. The starting citation can be obtained from a number of sources: a key paper already known to the inquirer, the list of references in that paper, or similar papers cited in monographs, encyclopedias, or other works. One must remember that it usually takes from 9 to 12 months before an article can be cited in the published literature and that additional time is required for it to be printed in the index. Citations also appear, however, to literature published 100 or more years ago, so any classical paper may also provide a good starting point for a citation search.

Citation indexes are effective in searches on methodology because these aspects are frequently ignored in conventional indexes, whereas authors often cite methodology papers in their lists of references to avoid detailed description (150:46). Methods and apparatus are frequently given the names of their originators who have described them in the literature and thus provide another means of entry into the *Citation Index*. These kinds of eponyms, personal names assigned to many phenomena in science (diseases, chemical reactions, anatomical structures, tests, etc.), can be used in the same way. Of course, eponyms can also be used in searching language-oriented indexes like *Index Medicus,* and they can be used in free text or natural language searches in online systems, as we will see later. However, not all eponymous diseases are indexed under the originator's name in *Index Medicus,* or they may be subsumed under broader headings. An example is the Gianotti-Crosti syndrome, an exanthematous eruption of the extremities which appears suddenly in young children. The key article which can be found by consulting a textbook or an eponym dictionary (230, 289) is:

> Crosti A, Gianotti F. Dermatose eruptione acrosituée l'origine problament virosque. *Dermatologia* 1957, 115:671–677.

MeSH indexes the syndrome under the general term *Acrodermatitis,* and articles on the syndrome must be retrieved from among many articles. The 1975–1979 *Cumulative SCI* lists four articles citing the Crosti article which are directly related. On the other hand, classical papers such as those associated with eponyms are sometimes no longer cited because they have been subjected to what has been called the "obliteration phenomenon" (164). This illustrates a primary maxim of literature searching: no single source is suitable for all purposes.

The printed version of the *Science Citation Index* can be searched in all the ways described above, but the computer-based or online version presents much greater flexibility and capabilities, as we shall see.

Citation Analysis

The availability of large machine-readable files of source articles and citations has stimulated information scientists, sociologists, and historians of science to use them for purposes other than information retrieval. They have been used to measure and evaluate the productivity of scientists and the journals in which they publish. For several years, the Institute for Scientific Information has been analyzing citation frequencies in and among the journals in its data files as means of monitoring and evaluating them. Since 1975, the results of these studies have been issued in a compendious volume each year called *Journal Citation Reports.* Citation frequencies are listed by journal title in alphabetic order, showing for each title the number of source articles it has published during the current year and each of the two preceding years, as well as the number of citations it has received for these periods. Duplicate references in the same source article are counted as single citations (150:149). This information is also listed by rank of citation frequency, showing for instance that for 1981 the most cited title was the *Journal of Biological Chemistry* with 117,001 citations and that the *Journal of Lymphology (Zeitschrift fur Lymphologie)* shared the 3,898th rank with fifty other titles as among the least cited titles.

Frequency of citation can serve to some extent as an objective measure of the significance of a journal title, but it can readily be seen that it may also be a function of the number of articles a journal publishes in a given year. Thus, older titles would tend to rank higher simply because they had been published longer. To equalize the citing potential of each journal, the number of citations received by any journal is divided by the number of source articles published in that period. The ratio derived, called the "impact factor," is defined as the "average citation rate of a journal article" (150:149). *Advances in Immunology,* which ranked 527 in the citation frequency list with 448 unique citations to a total of 14 articles published in 1979 and 1980, ranked first in the rank of impact factors with a ratio of 32.00, showing that each of the source articles was cited an average of 32 times in 1981. Almost three-fourths of the source journals had an impact factor of less than 1 in 1981, indicating that each of their source articles was cited on the average less than once, and that probably many of them were not cited at all.

Another measure, called the "immediacy index," is calculated to show how often on the average a source article is cited during the year in which it is published. Thus, *Advances in Immunology* with eleven citations in 1981 to five source articles published in 1981 has an immediacy index of 2.200, and the *Journal of Clinical Microbiology* with 93 citations in 1981 to 382 source articles published that year has an immediacy index of 0.243. This may be a function of such factors as publication lag time, but the immediacy index can also be related to obsolescence or what we have learned to call "use of the literature through time."

A more revealing measure is provided by two other sections of the *Journal Citation Reports.* The first is a "Cumulative chronological distribution of citations to cited journals" which indicates the percentage of total citations received in a particular year that are accounted for by articles published in the current year or in any one of the preceding 9 years. Thus, for the *American Journal of Physiology,* 2.46 percent of the citations it received in 1981 were for articles published in that year. The articles published in the preceding 5 years accounted for an additional 37.83 percent, and 56.9 percent of the citations covered the 10-year period. In a second listing which relates more directly to obsolescence, the journals are ranked by their "half-life," defined here as the number of years preceding the current year which account for 50 percent of the total citations received in a given year. Only those journals with 100 or more citations for the year are included. This half-life varies considerably from journal to journal. In 1981 it was 0.6 year for *Molecular and Cellular Biology* and 10.0 years for the *Yale Journal of Biology and Medicine,* a rank it shared with several hundred other journals.

The *Journal Citation Reports* also provides each year an analysis of the source journals by the frequency with which they are cited by other journals in that year *(Citing Journal Package)* and the frequency with which the articles in the source journals cite other journals *(Cited Journal Package)*. These data are presented in tabular form and show the distribution of citations for each of 10 years including the publication year. Thus, they serve both to define the relationship between journals and to help identify the core list of journals in any specialty. For example, only forty-three journals account for 85 percent of 13,824 citations received by *Virology* in 1981. All but six of these titles also show up on the list of journals which are cited by *Virology*. The top ten citing journals account for 67 percent of the citations to *Virology,* and 57 percent of the references in *Virology* are also accounted for by ten journals. These tables also supply information on the use of the literature through time. Thus in 1981, 82 percent of the articles in *Virology* cited in other journals were 10 years old or less. Analysis of the cited journal packages reveals that in many cases the most cited journals in any discipline are not necessarily those most closely associated with the field by title. Some of this information is useful in determining which journal titles an investigator needs to follow for current awareness.

The length of a candidate's bibliography has traditionally been among the criteria used in academic institutions by promotions and tenure and search committees to measure a candidate's contribution. It has been argued, however, that the number of times authors' papers are cited by their peers would provide a much more discriminating measure. Of the 40 million citations processed by the Institute for Scientific Information during a 15-year period, less than 1 percent were cited ten times or more during any one year (173). Sociologists for several decades have studied the relationships between scientists' productivity and the recognition of the quality of their work (81). It was not until the application of computer-based citation analysis to the problem that more objective measures could be achieved. One such measure can be obtained simply by counting the number of citations under any author's name in the *Citation Index.*

There is a natural tendency to be suspicious of any kind of quantitative measure of creative performance. A large number of studies, however, have shown strong positive correlations between citation rates and other indications of quality such as honors, awards, memberships in honorific societies, judgments by peers, and other forms of recognition. A study of the Nobel Prize winners in 1962 and 1963, for example, showed that they were cited thirty times more than the average citation rate for other scientists working in their field (150:64). The citations received by Nobel Laureates in physics from 1955 to 1965 averaged 58 in 1961, compared to an average of 5.5 for all other scientists cited in that year (81:379). The list of 300 most cited authors in the *Science Citation Index* from 1961 to 1975 included 42 Nobel Laureates, 110 members of the National Academy of Sciences, and 55 Fellows of the Royal Society of London (348:58).

The following case is not a typical example, but it illustrates how citation counts can be used both as a measure of quality of a scientist's productivity and as a qualitative filter. E. A. K. Alsabti published some sixty or so papers between 1977 and 1980, all of which have been alleged or shown to have been plagiarized from various sources and published in relatively obscure journals (57). The cumulated *Science Citation Index* for 1975–1979 lists thirty source articles as having been written by Alsabti. The *Citation Index* for that period lists only thirteen citations of these articles, and they are all self-citations. The citations in the 1980 *Science Citation Index* are also self-citations with the exception of several citations from letters and editorials which call attention to Alsabti's malfeasances. The difficulty of expunging fraudulent or plagiaristic articles from the literature, however, is demonstrated by the fact that the 1981 *Science Citation Index* contains several innocent citations to two of the Alsabti articles.

It is generally agreed that citation counts do not provide an ideal measure of the quality of research performance or of its significance. It is also argued, however, that they currently represent the best objective and quantitative indicator we have (80:xi). Other factors which affect citation rates have also been explored: the influence of the author's rank in the scientific community, the so-called "Matthew effect" (308), the variations in frequency with which work tends to get cited in different disciplines, the role of self-citation, and the fact that the *Citation Index* provides data only for first authors. It has been found, for instance, that mathematicians tend to cite half as many papers as biochemists (173). The influence of this factor is minimized because scientists are usually compared to others who work in the same discipline. The effect of collecting data only on first authors has been hotly debated, but it is reported that this "does not affect substantive conclusions" (80:33) about citation frequencies.

There is also evidence that self-citations do not significantly distort comparisons of citation frequencies. For one thing self-citation is a common practice in all disciplines. A certain air of opprobrium seems to attach itself to the term "self-citation" as if it were an objectionable form of self-aggrandizement. The reasons for citing your own work are as cogent as those for citing any other author's work, to connect the present work with previous work and to provide supporting evidence and clarification. Self-citation rates seem to vary from 10 to 20 percent from field to field. One investigator reviewed two subject areas, plant physiology, where she found the rate was 16.6 percent, and neurobiology, where it was 17.5 percent. In both fields articles which included no self-citations were less than 8 percent. The number of self-citations per paper ranged up to sixteen an article, with a median between two and three for the first subject and between three and four for the latter. The author concluded: "Self-citing, then, is a common and fundamental attribute of scientific articles, and has a function which essentially is not different from that of other forms of citing" (428).

The conventional wisdom is that eminent scientists tend to get much greater credit for their contributions than lesser known scientists. Studies of the relationship between the recognition and status of scientific authors and the acceptance/rejection ratio of their articles have demonstrated, however, that rank exerts little influence.(484). The issue is even more difficult to resolve with citation rates and for some of the same reasons. First, it is generally agreed not only that eminent scientists tend to publish more but also that they tend to publish better papers. It has also been demonstrated, as I pointed out earlier, that citation rates correlate highly with such factors as scientific eminence and peer recognition, but the only way to assess the effect of these factors would be to compare these data with citation rates in which all the cited authors were anonymous, a difficult objective to achieve.

There are other ways in which citation analysis has been used in addition to those I have already mentioned. Merton calls it "a research tool that is largely specific to the history and sociology of science" (311:49). It, for instance, played an important role in the new discipline of scientometrics, defined as "the study of the measurement of scientific and technological progress" (167). It has been used to provide insights into changes in emphases in scientific research and to identify seminal papers. Scientific ideas have lineages which can be traced back through references to earlier documents. Citation sequences can serve like genealogical charts in which new documents occupy space as progeny of the intermarriages of ideas represented by older documents. Price sees the citation process not as one of new papers referring to old papers but as old papers generating new papers (368). An interesting new tool in helping to chart and map these relationships is represented by a technique called co-citation analysis. Not only does it help to make the invisible college visible; it also helps to define and provide a name for the area of their common concern (150:98).

Co-Citation Indexing and Analysis

Co-citation indexing and analysis is based on the hypothesis that two documents (articles) which are cited by another document are more likely to be related in subject matter than those which are not cited together (412). The citing paper or source document thus establishes links between the papers it cites. These links can be analyzed by sophisticated computer programs to provide data which can be used to develop "maps" of scientific specialties, which can also serve as an auxiliary tool in literature searching. The process has been described extensively (413), but essentially it involves a statistical technique called "clustering" which is used to identify resemblances between members of a population based on variables in that population (153). In the adaptation used by the Institute for Scientific Information, the population is the total number of source documents included in any year in the *Science Citation Index* and the variables are all the papers they cite (citations). Out of the over 7.5 million citations which appear in any year, only articles cited a given "frequency threshold" are selected. This may vary from discipline to discipline. Setting the threshold at seventeen times a document, for example, reduces the citations for analysis to fewer than 25,000 (153). These citations are processed again to select for clustering those which exceed a given co-citation "strength threshold," defined as the proportion on which a given pair of citations are cited together. These relationships can be mapped graphically to provide a cluster which defines an area of research more effectively than trying to fit it into established categories.

These "clusters of highly interactive documents" are also called "research-front specialties" (225). Some 7,942 clusters have been identified from the 1979–1981 literature for the 1982 *Science Citation Index*. The specialties are defined by the core papers which make up the citation cluster, but for purposes of identification they are given names such as "Active transport in bacteria," "Cell response to hyperthermia," and "Histamine-receptors and immunity" which are derived from the papers in that particular cluster.

These data are interesting and useful for studying the historical development of a research problem and for understanding the structure of science, but they also provide a useful tool for organizing and retrieving information. A literature search based on a "research front" can be said essentially to be a search on all the most highly cited papers in a particular specialty. These can be used in search strategies in lieu of single citations, to provide groups of articles which are cited together, as well as current articles which cite two or more articles in the cluster. In 1982, the *Citation Index* section of the *Index to Scientific Reviews* was replaced with a *Research Front Specialty Index* to provide more effective access to the review literature.

Co-cited authors, that is, authors who are cited together in the same paper, have also been used as a means of subject mapping as well as an information retrieval strategy (463).

These kinds of searches are easy in SCISEARCH, but they can also be performed manually in the printed citation indexes by coordinating cited authors to locate papers which cite them together. The process is greatly facilitated in the online system, where a command to search on both Jones and Smith as cited authors would deliver a list of source articles in which both of these authors are cited. Co-citation analysis has also been seen as a method of producing state-of-the-art reviews. The first prototype segment was published in 1981 under the title *ISI Atlas of Science: Biochemistry and Molecular Biology* to which I referred in Chapter 4 on secondary sources.

As I indicated earlier, questions have been raised about some of the assumptions which underlie citation indexing as a technique for retrieval and evaluation of the literature. Reservations have been expressed also by the producers of these indexes, who have cautioned from the very beginning against the "possible promiscuous and careless use of quantitative citation data for sociological evaluation, including personnel and fellowship selection" (149). They also recognize that scientific merit is not always the only reason an article is cited and that there are journals which are useful and significant for other purposes than yielding citations. Citing a paper does not necessarily imply that it has been read, nor are all papers which are read cited, as Price pointed out (368). The *Citation Index* does not provide access to half of the papers included in the *Source Index,* since they may never be cited, although they are available through the *Permuterm Index.* Citation analysis implies an equivalence among the units counted and tends to weigh a paper published in a high-quality journal with one that appears in a journal of lesser standing.

These problems have been discussed extensively in the literature (80, 177:61, 297, 414). Nevertheless, a number of studies show that there is a great deal of validity to citation indexing and analysis. It has been demonstrated that authors who are more frequently cited are more likely to be recognized by their peers.

In a 1969 study, selected authors of scientific articles were sent copies of papers which cited their work and were asked to judge how relevant they were. They indicated that 72 percent of the references were closely or directly related and that only 4.9 percent were not related (22). In the past two decades, citation analysis has proved itself an important and powerful tool both as means of information retrieval and as an aid in understanding and evaluating the scientific literature.

Chapter 10

Searching the Literature

When we speak of searching the literature today we are no longer referring solely to printed sources but also to electronic ones, or what currently may be more accurately termed print sources available in electronic form. These sources range from numerical data tables to full texts of journals, of which many are currently available and in use. Although electronic searching is becoming more and more prevalent, there seems to be no question that printed sources will coexist with electronic for some time. Electronic databases may be more efficient in many cases in relation to the volume of results and the expenditure of time, but they are not always the best solution to every information problem. Much of what I have said about scope, coverage, and search techniques for printed sources is also relevant to their electronic counterparts.

Everyone at one time or another has probably had a similar experience in searching an information source. It does not yet have a name like *déjà vu,* and perhaps it can best be described by the statement: "I know it's there but where did they put it?" We sense somehow that the information we are seeking must exist in the source we are searching, but are unable to find the key or association trail which will lead us to it. The answer to the question lies not only in the consideration of the logical implications of the problem (where does it fit in the scheme of things?), but also in the state of the art relating to the subject of inquiry, the predilections and interests of the individuals responsible for assigning it a place, and to a large extent the composition and the characteristics of the database in which it may be located.

A literature search problem resembles in many ways any kind of research problem. It is more distinctly a "search" problem rather than a "research" problem because we are concerned not with acquiring new information, as we are in research, but with recovering or locating already existing information. The two processes, however, can be equally creative and imaginative, because they both involve the recognition of relationships, particularly when they are new relationships. Success in either case depends to a large extent on the ability to state the question in researchable terms. In information systems that means to phrase it in a way which is consonant with the source you are using. This is particularly important when a second party such as a "search analyst" is enlisted in the task. You must be sure that you are asking a researchable question, that is, that you have the necessary concepts and tools to attempt an answer. Medawar in one of his essays on the scientific enterprise stated (300:87):

> No scientist is admired for failing in the attempt to solve problems that lie beyond his competence. The most he can hope for is the kindly contempt earned by the Utopian politician. If politics is the art of the possible, research is surely the art of the soluble. Both are immensely practical-minded affairs.

The question of "solubility" or solvability may, however, be most difficult to answer in advance of making the attempt. ". . . [I]t is unrealistic in general," says one information scientist, "to ask the user of an IR [Information Retrieval] system to say exactly what it is that she/he needs to know, since it is just the lack of that knowledge which has brought her/him to the system in the first place" (27).

Formulating Search Strategies

To avoid the type of searching that engineers sometimes refer to in describing the erratic vacillations of the indicator needle of an electronic measuring instrument, it is wise to plan your literature search before you begin it. This is particularly true in planning online database searching where the meter keeps ticking relentlessly, adding to your costs of accessing the service. To some extent, it does not matter whether the search plan is for an online or a manual system. As we shall see, however, one must be concise and explicit about computer-based search strategies, whereas in a manual search some of the search parameters do not need to be eternalized or made explicit. In either case the time spent in planning will help in conceptualizing the problem. This idea is reflected in what was sometimes called the "black box" theory in the days when the National Science Foundation was more actively funding the procurement of computer equipment. The feeling was that the time spent in planning the program to be implemented by a computer was not wasted even if the computer was not awarded, because it enabled the planners to see the problem more clearly and to carry out the solution more efficiently even if they had to do it manually. The decision to use either printed or online information sources will, in any case, be one outcome of the search planning process.

As a part of planning the search strategy you must also decide how much time is available for the activity and what degree of comprehensiveness or completeness must be obtained. In many cases the sources will be manifold. Despite the overlap and redundancy that appear in the sources, searching other sources will almost always add new information. Bradford's law of scattering applies here because it tells us that a 100 percent increase in effort seldom results in a 100 percent increase in results. Doubling your search time may only increase the number of references retrieved by 25 percent. This may not be very significant if you have already retrieved more references than you need for your purpose. Doubling your efforts again may only increase your returns by 5 percent. At this point you may wish to invoke the Law of Diminishing Returns, a general law that applies in many situations.

There is no standard formula or mechanism that you can apply to finding information, says Ziman. ". . . [T]he choice of strategy will depend on the particular circumstances" (477:315). Formulations will vary from searcher to searcher even when they are searching the same database and using the same system. It is therefore not useful at this point to discuss specific search strategies since they are specific to particular searches and to particular searchers. As with most activities, however, it is useful to follow a general procedural plan. Some steps which have been suggested are:

1. To conceptualize your need for information in your own terms as clearly and precisely as possible on the most specific level you can. Indexers who use controlled vocabularies are instructed to use the most specific terms available to them to describe the concepts in the document they are indexing. Even if the database you choose to search may not contain the terms at that level, they may lead to more generic terms which are in the system.
2. To decide on the scope of the search in terms of comprehensiveness, period of time to be covered, and languages to include or exclude.
3. To select what seems to be the most appropriate source to search, depending on what you know about its subject scope, coverage of the literature, and other features, remembering that your goal usually is that of achieving the highest yield with the least effort.
4. To translate the terms you have selected into those of the source or system you plan to use, in a form which is amenable to the capabilities of the system. Indexers select those terms which they think most succinctly describe the document and which they think will lead an inquirer to the information contained in that source. Assignment of the appropriate terms will depend on the knowledge of the indexers and their ability to anticipate what terms the inquirers will select and what those terms will mean to them. Thesauri may not always be able to lead you to the approved term. It may be included in a more generic term or entered in a reversed order, *Radioautography* instead of *Autoradiography,* or it may be part of an inverted term, *Surgery, Plastic* instead of *Plastic Surgery.* It may use different roots with the same meaning, *Cerebri-* or *Enceph-* for *brain* terms, or it may be precoordinated with other terms as in *Laryngeal Neoplasms.*
5. To decide which are the major and the minor concepts in your inquiry, and which terms fall into the separate groups to be coordinated.
6. To test your strategy on the system, revising it as necessary if it does not produce acceptable results. Online systems are described as "interactive" because they permit you to see the results of your search strategy at once

and to narrow or broaden it as necessary. You can, of course, interact with manual systems also, but the interaction is of quite a different character.

It is also important to be aware of the history of a term in a system, particularly in those with controlled vocabularies. Be aware whether more generic or specific terms were previously used for the concept or were used in a different form, as when *Down's Syndrome* was substituted for *Mongolism* in MeSH. In computer files inquiries under the old term may be referred to the new one. This help is not always provided in printed indexes, although each annual issue of MeSH supplies a list of the terms which have been added and changed and in the contents notes when earlier forms were used. If a document relevant to the inquiry is known to the searcher, it may help to refine the search strategy. Online systems often permit you to display the terms under which an article has been indexed when it is retrieved. The title words in a relevant article may also supply additional cues.

Hawkins (211) describes three different types of search strategies which, although they relate particularly to online searching, can be applied to manual searching as well.

1. The building block strategy in which each part of the inquiry is conducted as a separate search and the results are combined—a procedure, he says, which must be carefully preplanned.
2. The successive fraction strategy in which one starts with the term which will produce the largest number of references and then adds specific concepts to narrow the search. This strategy is useful, he says, when the topic is vague or broad and when useful restrictions can be added.
3. The citation pearl growing strategy referred to above in which a known relevant citation is examined to determine how it was indexed.

Why do search formulations fail? Some reasons are the use of inappropriate vocabulary terms, searching on either too specific or too general a level depending on the nature of the search, and failure to cover all reasonable approaches (124). I shall discuss some of these problems when I deal with online systems, but they are often applicable to searching printed indexes as well.

Online Search Systems

The term "database" has been more or less appropriated by information specialists to characterize a discrete collection of machine-readable information stored in a computer. It can, however, also be used in a generic sense to describe any collection of information, whether stored in handbooks, indexes, dictionaries, or any other printed information source. The term "document" can also be used in a generic sense to describe a separate information unit, whether it is a book, a journal article, or a patent specification. Databases vary as to the kind of documents from which they extract their information, as well as to coverage, format, etc., as we learned in our discussion of primary and secondary sources. Many of these are now available in electronic or online as well as printed forms, but the choice of an appropriate online database still depends on the same factors that guide us in the choice of a printed one.

Many of the online databases in current use came into existence when computer-based photocomposition was introduced to produce the printed version of such publications as *Biological Abstracts, Chemical Abstracts,* and *Index Medicus.* The same data then could be made available for searching on remote terminals by various computer programs. The recent phenomenal growth of online searching has also been made possible by the development of time-sharing computers that enable many individuals to use the same computer at the same time. This development was accompanied by the enhancement of telecommunications capabilities which can connect terminals with computers anywhere in the world.

There are four primary components of most online database services. First, there is the database producer who collects the data, organizes it, indexes it, and formats it in electronic form. These include the producers of the major indexes I have discussed as well as a host of others. In many instances, the database producers do not make them directly available from their own computers. Second, there are the vendors or suppliers who may acquire several of these databases, store them in their own computers, and provide access to them for a fee. They also develop and provide the search protocols or procedures that are required to retrieve the data at remote terminals. There are three major vendors of database services of this kind in this country: Bibliographic Retrieval Services (BRS); DIALOG, a subsidiary of the Lockheed Company; and ORBIT, supplied by the Systems Development Corporation. Collectively, they offer access to well over 200 databases, many of which are offered by all three vendors.

A third component of the online service is the telecommunications network which provides the connection between the vendor's computer and the user's terminal via telephone lines, cables, satellites, etc. There are a number of these networks like TELENET and TYMNET. They provide local telephone numbers through which a user's terminal can be connected with a remote computer. Some database producers, for example, Chemical Abstracts Service and the National Library of Medicine, also provide users with direct access to their computers through these networks and thus also act as vendors. Some online databases can be accessed directly by placing long-distance calls which connect with ports in the host computer. It is usually considered less expensive, however, to route a call through the nearest city which has a network number (node) because the distance from the node to the host computer is not a factor in computing network costs.

There is a wide and dizzying variety of access equipment or terminals ranging from print terminals which look like typewriters and may soon be produced in pocket-sized versions to elaborate cathode ray tube (CRT) devices which provide visual displays of the information and can be equipped with many different kinds of keyboards, peripheral printers, and electronic storage devices.

Some of these terminals may be "dumb," or capable of only transmitting or displaying data; others may be "intelligent," with the capability of storing and editing the information retrieved. Microcomputers or small personal computers which are capable of other applications can also be modified to access online databases. With appropriate programs they can also "download" or store the information received on peripheral devices such as floppy or hard disk drives. For most terminals, unless they are capable of direct phone connections, some sort of modem or acoustic coupler is required to "modulate" the signals produced by the terminal or computer into a form which can be transmitted over telephone lines.

Terminals operate at various output speeds, i.e., the rate at which the information is displayed on CRTs or printers. This speed is designated as the terminal's "baud" rate. Until recently, many terminals were operating at 300 baud. More and more today operate at 1,200 baud or at even higher baud rates. This is a significant factor because costs of online service tend to vary according to the time spent online or in direct access to the remote computer. Vendors, of course, are aware of this, and alternative methods of pricing are being introduced.

The vendor controls access to the databases by making clients identify themselves online through a "password" or code designation which is assigned when the requester applies and sets up an account on which to be billed. The password is known only to the client and the vendor and is not displayed when the system is accessed. The costs are usually collected by the vendor, but they include the royalties paid to the database producer which may vary considerably from database to database. They also include the vendor's own charges for providing access and, finally, the telecommunications costs. As an example, a 15-minute search which on some databases may cost about $16 might break down this way: $6 to the database producer at $24 an hour, $8 to the vendor at $32 an hour, and $2 to the telecommunications services at $8 an hour. Other miscellaneous charges may be included depending on whether the user wants some of the results printed "offline," that is, by the vendor's computer, and then mailed in printed form, an option which is usually considered when the output is exceptionally long.

A few libraries, particularly some industrial and nonacademic research libraries, do not charge their clients for online searches, recognizing that they substitute for manual searches for which traditionally no charges may have been assessed. Those libraries which do charge generally attempt to recover only their direct costs, which are the same as an individual would pay for conducting the search without assistance. They usually do not charge for overhead costs, amortization of the equipment, and the extensive training and updating that is frequently required to access some databases. A study of the utilization of databases in 1979 showed that charges to users varied considerably depending on the environment in which the search took place, the cost of the database accessed, and the amount of time spent online at the terminal, which also could vary from one searcher to another (246:157).

Despite dramatic reductions in the costs of storing information in computers, there have been no comparable reductions in telecommunications costs or in the costs of producing databases, which, on the contrary, have tended to rise. Vendors and database producers, becoming concerned about the ease of downloading and reusing the data electronically, and about the substitution of online services for their print products which up to now have borne a considerable part of the costs of maintaining their systems, are beginning to change their pricing policies.

The diversity of vendors and of databases is accompanied by a diversity of search techniques, i.e., instructions or "commands" (as they are sometimes peremptorily called) which enable the computer to accomplish the search objectives. This leads us to the fourth "component" of online database services, the "searcher," who may be an individual who is especially trained to use these systems or the individual inquirer who has made the effort to learn how to use them. Before I discuss this as an issue, however, let us consider the circumstances under which an inquirer will choose to use an online service over a printed source, the extent of existing online databases, and the search techniques that are used in online systems.

Online vs. Manual Systems

Ziman's statement that "No machine can replace the searching *eye* linked to the discriminating brain" was written in 1972 (477:318). The technology and the capabilities of machine searching have changed considerably since that date, but the comment still retains a certain validity. Online searches require the machine to make exact matches on the combination of characters the searcher has chosen to formulate the search. Some programs provide for spelling errors and for "truncation" (accepting words of which only a part such as the stem has been entered). Most computer systems, however, are extremely recalcitrant. If you don't do exactly what they require you to do—for instance, if you enter a comma where a period is indicated—they won't respond appropriately. The logic of an information search is not essentially different when it is performed in manual or machine form. However, in the machine form the logic and the mechanisms are highly visible and must be articulated in the process of the search, while in manual systems one can more easily test and add term coordinates as individual terms are searched, a useful advantage particularly when all term coordinates cannot be specified in advance.

Some searches can be more readily performed in a printed index, particularly when only a few references are required, or when the inquiry is well defined by a single search term which is available in the system. Another reason for using manual systems is that most online databases are extremely limited in terms of the time span covered. *Chemical Abstracts* online (CASEARCH) goes back only to 1967, whereas the print version covers the literature back to 1907. MEDLINE covers the literature only from 1966, whereas the printed indexes of the National Library of Medicine take you back to the beginning of printing. Online systems cut you off effectively from most of the literature of the first two-thirds of the twentieth century as well as from the great retrospective bibliographies such as the *Catalogue of Scientific Papers* of the Royal Society of London, which covers the scientific literature of the nineteenth century. This may not be as important in the life sciences, where the emphasis is primarily on the current literature, but total reliance on computerized databases also carries with it some risks. Important papers may be missed, sometimes leading to the repetition of research that has already been reported in the literature. It is not often that an investigator is moved to exclaim that a report ". . . missed an article of mine from 1966 in which I had solved completely one of the topics they addressed," but it must occur more frequently than we can guess (351).

Another disadvantage of online search services is the need for a search intermediary. This may require explaining in detail the nature and goal of the search, which may not be easy to articulate at this stage even for the investigator. It may just as easily be argued, however, that the need to carefully articulate and explain the object of the search may be an advantage in clarifying it in the investigator's own mind.

Computer-based searching, nevertheless, has great advantages and power. The decision to use manual over machine systems is sometimes made on the basis of costs, but online searching can be much more economical, particularly when the cost of time expended enters into the calculations. Online searching provides the ability to respond to questions that would have been extremely difficult if not impossible to carry out with traditional methods. The computer not only can search many concepts simultaneously, but also can include other facets in the search, such as publication dates, authors, and language. In manual searches you usually select one key term (a good strategy in either manual or machine systems is to select that term under which the least number of items will be displayed) and then depend on the other terms in the citation to help you determine whether it is relevant. In computer-assisted searches, the inquirer can also coordinate with all the other descriptors, tags, and headings which have been assigned to that item but are not displayed in the manual system.

Items in online systems are stored in random order and in that sense are self-cumulating. This may eliminate the need to search long series of successive volumes to cover several years of the database. One year of the *Cumulated Index Medicus* today covers almost 3 feet of the shelf space. Searching 5 years of the file may provide beneficial physical exercise but can also be time consuming.

The issues relating to controlled versus natural language indexing systems change radically when they are confronted in machine systems, as I indicated earlier. Natural language "textwords" (words from titles or abstracts) can be used in conjunction with words specified by a thesaurus or controlled vocabulary. *Index Medicus* and MEDLINE both use the same vocabulary but with important differences. In *Index Medicus* the average number of subjects assigned to an article is three, while in MEDLINE the average is twelve. If the same number of subjects were covered in the printed version as in the online version, it would average about 48,000 pages a year instead of the usual 7,000. In addition to the *Index Medicus* citations, MEDLINE also includes the entries in the *Index to Dental Literature* and *International Nursing Index.*

The indexing language for MEDLINE is also much richer since it includes the 5,000 "minor descriptors" (subject headings) which are not included among the 10,000 major descriptors available for *Index Medicus.* Subheadings can also be used directly in the search formulation. These do not include the other descriptors such as the "check tags" (descriptors relating to age of subject, experimental animal, sex, etc.) which bring the total to about 25,000 terms available in MEDLINE to characterize the content of an article. In MEDLINE subject headings are tagged to indicate those which represent the major emphasis of an article. Searches can be restricted to major emphasis terms and help to increase the relevancy of the items retrieved.

Computers also make it possible to use the advantages of both the alphabetical and classified systems at the same time by coordinating a specific subject with all the aspects of a general subject. You can, for example, in MEDLINE coordinate *Diabetes mellitus* with any aspect of *Nutrition* or *Diet* by "exploding" the latter terms, that is, instructing the computer to include any of the terms subsumed under these terms in the search. In the *Excerpta Medica* system the database can be searched simultaneously by EMCLASS number assigned to every level of 3,500 hierarchically arranged terms. They are classed on four levels of generality, along with words in MALIMET, which is *Excerpta Medica's* controlled vocabulary. In BIOSIS keywords can be searched in conjunction with its classified "Concept Codes" and "Biosystematic Codes" at any desired level of specificity or generality, a task which is quite difficult in the printed version.

Another advantage of online systems is that the information is frequently available in a more timely manner than with their printed versions. One study, for instance, checked the dates between the arrival time in the library of certain printed indexes and their availability in online systems and found that for *Index Medicus* and MEDLINE the difference was 43 days. This is not true of all printed indexes because the same study indicated that, although the time lag between publication and availability in indexes was 129 days for *Index Medicus* and 86 days for MEDLINE, it was only 31 days for the printed *Current Contents* (366).

In the early days of computer-based searching, users were cautioned not to use the computer as a printing device. It is, however, actually a superb printing device, as word processing and computer-based photocomposition system have demonstrated. The warning pertained more perhaps to using the ponderous power of the computer for simple lookups rather than for printing the results of a search when obtained, because surely having a machine automatically and accurately transcribing a citation is far superior to copying it by hand. The computer furthermore offers the capability of printing items selectively, printing them in various formats, and editing them after they are received or stored electronically in another file.

Online systems also have the ability to display the terms under which any document has been indexed, which can be a great aid in formulating a search if a single relevant article is already known to the inquirer. Once a search strategy for an online system is formulated, it can be revised and modified while it is in process. If too many citations are retrieved, the search can be narrowed by adding parameters or extended by increasing the generality of the search. In this sense one of the great advantages of the online system is that it is "interactive"; that is, it provides you immediate feedback on the results of your search. Finally, search strategies can be saved in online systems so that they can be repeated or modified when the search is updated, or processed at regular intervals to provide the kind of selective dissemination of information (SDI) services I discussed earlier.

Online Databases

In his *World Brain* published in 1938 H. G. Wells (461) advocated a central storehouse of easily accessible information which was constantly updated and which defined the world's state of knowledge at that moment. This fantasy may never be achieved because of the political and intellectual problems involved, but the proliferation of online databases is gradually moving us in this direction. In 1976, it was estimated that there were 300 online databases available (246:151) of which a vast majority were related to science, and of which 28 were specifically related to the life sciences. An online directory in 1979 listed over 500 such databases containing a total of over 70 million citations (465). A 1983 directory indicated that there were over 1,600 databases available online through 225 different vendors and covering the subjects alphabetically from "Accounting" to "Wines" (114), with new ones constantly being added. In November 1982, for instance, one of the standard biological abstract services, the *Zoological Record*, became available online as a joint production of BIOSIS and the Zoological Society of London. It corresponds closely to the print version in which approximately 6,000 journals are screened for articles of zoological interest. It began with 60,000 records from 1978, with bimonthly updates of about 10,000 records to begin in January 1983 (340). The number of records available in all systems has gone up accordingly. In MEDLINE alone the number has gone up from 2.5 million covering the literature from 1966 to 1976, to over 4 million to June 1983, and this does not include the records carried in the many other online databases maintained by the National Library of Medicine. A comparable increase has taken place in the amount of online searching. A report issued in 1979 predicted, "On-line searches are expected to increase to 3.7 million in 1985" (246:301). A footnote in the same report indicated that the number had already been surpassed by 1979. In the fiscal year 1981 there were almost 1.5 million searches carried out on MEDLINE alone (324).

Up to this point I have cited primarily online databases which are bibliographical, those which contain only references and in some cases abstracts. Bibliographic databases generally do not supply "data" unless they are included in the abstracts. They contain "citations" which lead you to the source of the data in journal articles, books, and other documents. There are, however, other kinds of databases now available online. They can be characterized as "factual" databases which provide information found in biographical and institutional directories, or in handbooks and dictionaires such as the ones discussed earlier.

There are also "numeric" databases such as those which provide quotations on securities from American stock exchanges, although ones of more scientific interest are being added. Recently, one of the online vendors (BRS) announced an online index, called SUPERINDEX, to major data and reference handbooks in science, engineering, and medicine which contained over 2 million of what they called "back of the book" index entries. In addition to these types of online databases, there are the so called "full-text" databases such as one which provides access to the entire contents of the *Encyclopedia Britannica*. A few scientific journals are already available in this form, and there are predictions that it will become a major form of scientific journal distribution.

These non-bibliographic databases are too numerous to list, but among those which are of interest to the life sciences are the following. The *Toxicology Data Bank* (TDB) is a textual numeric database produced by the National Library of Medicine which contains chemical, pharmacological, and physical properties, shelf-life, and toxicology data on over 3,500 substances of known potential toxicity. The *Registry of Toxic Effects of Chemical Substances* (RTECS), also produced by the National Library of Medicine, contains information on acute toxicity values for about 36,000 substances as well as some data on their carcinogenicity (249). CHEMLINE, which is a chemical dictionary online, contains 900,000 names for chemical substances and provides chemical abstracts registry numbers, molecular formulas, preferred nomenclature, and generic and trivial names for 450,000 chemical compounds.

Online Searching

There is not enough space here to consider the different search options, techniques, and protocols that are represented even by the databases that are most important to the life sciences. There are, however, a number of recently published manuals on the subject of online searching (207, 350). In addition, many vendors such as DIALOG, BRS, and ORBIT publish manuals which instruct users on the capabilities and command language to access the various databases they supply. Database producers like BIOSIS, CASEARCH, SCI-SEARCH, MEDLINE, and many others also provide manuals to facilitate the use of their system, whether they provide direct access or access only through vendors. It is quite likely, in any case, that much of what I might say about specific aspects of each system may be out of date by the time you read it. There are, nevertheless, certain standards and common procedures which remain relatively stable although vendors and databases may have different conventions regarding spelling, word order, and spacing, and different commands to carry out the same function.

Although the form of user cues and commands may differ, most bibliographic databases permit searching by a variety of similar access points including: title words, controlled vocabulary or thesaurus terms (when available), specific terms in a single category or all terms included in a generic category, author, journal title, and date or the period the user wishes to cover in the search. These may all be combined in a variety of ways in the search formulation. A combination of an author's name with a subject term, for example, will retrieve everything in the database in which the name is entered as one of the authors of an article on that subject.

Many of the electronic search techniques have been derived from or modeled after manual approaches. Textword searching for instance is an enhanced version of what we have called "keyword in context." Textword searching in online systems retrieves all the articles which contain the designated word in the title or in the abstract if included in the database, which indexers may not have considered important enough to index. The problems are similar to those of any uncontrolled vocabulary: synonymous terms and spelling variants for related concepts such as sulphur, sulfurous, sulphides. The problem can be dealt with in online systems by including all the variants in the search formulation. Online systems provide another aid, as I indicated earlier, which can be used in this situation, called "truncation." This may occur in several forms. For example, right truncation searches all terms or textwords with the same beginning characters: *Epilep* . . . will retrieve *Epilepsy, Epilepsies, Epileptic, Epileptics,* etc. Left truncation retrieves all words which end with the same characters: . . . *Otomy* will cover the names of all surgical procedures which end with this suffix. There is even a form of truncation in the middle, called "infix" truncation, in which only the beginning and the end of a word is designated but not the characters in the middle, a technique sometimes useful with chemical compounds such as *Tri* . . . *coboltate* (270:21). Some systems even permit you to designate the position of a selected full or truncated word in a particular context, either before or after another word, or within a designated distance from another word.

In virtually all online search systems these elements can be combined by logical operators derived from Boolean algebra. They deal with the mathematical relationships between statements by means of conjunction (AND), disjunction (OR), and negation (NOT) (185:37). Using AND in a search statement means that the computer must match all terms and other elements (year, journal title, etc.) conjoined in order to retrieve it. Each term added thus tends to reduce the number of citations retrieved. Search analysts, therefore, sometimes caution against "ANDing" too many terms and thus "ANDing yourself out." For example, JONES *AND* JOURNAL OF BACTERIOLOGY *AND* 1980 would retrieve the articles by Jones in that journal which are in the database, but only if they were published in 1980. The use of AND *narrows* a search by subtracting all the items which do not match any one of the terms combined in this way and means that both terms must be present in the citation for it to be retrieved.

The use of the logical operator OR indicates that either element so joined may be present in the citation. In other words the use of OR usually *broadens* the search by increasing the number of records retrieved. It is useful when synonyms must be included in textword searches or when categories must be added to a search statement.

Finally, the Boolean operator NOT (AND NOT in some systems) *excludes* the terms so conjoined from the search; that is, even if the citation meets all the criteria specified by the use of AND and OR, it will not be retrieved if it contains the term which has been negated. Asking for citations which contain the terms BLUE *OR* GREEN *OR* VIOLET *AND NOT* RED will eliminate all the items which contain the terms BLUE, GREEN, or VIOLET if they also contain the term RED.

The search programs follow these instructions in a priority order similar to those applied in computer arithmetic calculations: (i) division (NOT), (ii) subtraction (OR), and (iii) addition (AND) (349). The search statement CAT *OR* DOGS *AND* SURGERY will retrieve all articles on dog surgery and all articles on cats, but no articles on cat surgery, unless the subject was covered in articles which also dealt with dog surgery. Some search programs allow for nesting terms within parentheses which instruct the computer to operate on them as single statements. Thus, the above statement could also have been written (CAT *OR* DOGS) *AND* SURGERY to retrieve all articles on cat and dog surgery.

This search could also have been entered as successive statements which are then combined, to achieve the same results:

1. CAT *OR* DOG
2. 1 and SURGERY

The system would have responded with the number of "postings," that is, the number of items in the database which satisfy the designated criteria in each statement. In this way the database can be sampled to determine the number of items retrieved by a term or combination of terms. A common search strategy is to progressively build up sets of retrieved items and then to combine them with Boolean operators.

Search strategies vary from vendor to vendor and database to database. Alternative strategies can also be followed in the same database on the same search. Some general principles, however, have been suggested. Concepts and other search categories can be searched at varying levels of generality (exhaustivity) or specificity. It is considered good search strategy to increase the specificity or generality of only one search element at a time and to hold the others constant when you wish to narrow or broaden a search.

These few remarks do not, of course, exhaust the capabilities of online search strategies. They do, however, demonstrate the highly interactive nature of these systems. Not all online searches can be successful. Sometimes the database does not contain the information required. Sometimes it is difficult to develop an appropriate strategy which will retrieve the required information.

And sometimes a manual search in a printed index or handbook can prove a more effective and successful method of achieving the desired results.

Searching Directly or with an Intermediary

None of the procedures I have described is so complex that it cannot be learned by any investigator. Indeed, some investigators who are not by training search analysts do conduct their own searches. They usually need to acquire their own passwords from the vendor to establish themselves as bonafide and billable clients. There is, however, some controversy about whether investigators should do their own searches or use an intermediary, a "searcher" or "search analyst." In a survey of online bibliographic retrieval services conducted in 1979 it was discovered that less than 1 percent of searches were conducted by what are called "end users" (462). Most users do not have the time to learn the necessary techniques, nor do they search frequently enough to remember them or to keep up with system changes. Systems can be used on a simplistic or highly sophisticated level, and some search formulations require elaborate combinations of search terms.

Online searching is an interactive process in which the search can be modified as it proceeds, and optimum conditions are present when the requester is involved while the search is taking place, so that search results can be evaluated and modified as necessary. Studies have shown that searches are carried out at a higher level of precision (fewer irrelevant references) and retrieve a larger number of the relevant references when the requester is present (312). However, the 1979 survey cited above indicated that this occurs infrequently in online searching and that "eighty percent of all searches are processed by a searcher who works at the terminal without the user present . . ." (462:5).

Successful searches can be carried out in response to telephone or written requests, and when the requester is not present, but they depend on the ability of the requester to *articulate the needs* to the searcher and for the searcher to *clearly understand them*. This is sometimes difficult with technical topics with which the search analyst may not be familiar. Inquirers usually phrase the query in terms of their assumptions about the capabilities and limitations of the system, which may not be realistic. It is highly desirable, therefore, for the requester to spend as much time as possible negotiating the search with the analyst, defining the problem, and reaching agreements on the selection of access points and search strategies.

This situation may change in the future because there are more "user friendly" systems in development, and some universal search programs are already available in which the databases offered by a single vendor may be searched with the same protocols. Among these are "BRS After Dark," "Knowledge Index" by DIALOG, and SCI-MATE.

Evaluation

The most cost-effective literature searches are those that produce the maximum number of relevant items at the least expenditure of time and money. Two primary measures have been proposed for assessing their efficiency. "Precision" is a measure of the percentage of the references retrieved which are considered germane to the inquiry. If a search retrieves ten references and five of them are judged to be relevant, the precision ratio is 50 percent. The other ratio is a measure of "recall," the number of relevant documents retrieved in relation to the total number of relevant documents which are known to exist in the database. If a search retrieves five relevant documents and it is known that there are actually ten in the database which satisfy the needs of the inquiry, the recall ratio is also 50 percent. Precision estimates are relatively easy to make because they are based on examining the actual results of the search. Recall estimates are more difficult because it is necessary to know how many relevant items there are in the database, and this must be determined by methods other than those used in the search strategy being evaluated. Ideally, one should visually examine every item in the database to determine whether or not it is relevant but this is virtually impossible with large databases. Each measure is relatively meaningless in isolation. Recall has little significance unless it is related to precision because high recall ratios can easily be achieved by broadening the search enough to retrieve a large number of the relevant items, even though they represent a small percentage of the total retrieved.

Recall and precision ratios vary considerably from search to search. This may not be significant when only a few references are retrieved or only a few are required so that one can easily ignore the irrelevant ones. Even zero results are sometimes acceptable especially when an investigator prefers to discover that no one has reported on the line of research which is being initiated.

The first large-scale evaluation of a major computer-based information system was that reported on the MEDLARS system before it went online with MEDLINE, when searches were still being performed in batch mode, that is, at a central computer and then mailed or otherwise delivered to a requestor (268). The results of that survey showed that the system was on the average operating at a 57.7 percent recall and a 50.4 percent precision ratio.

> That is, on the average, over the 299 test searches, MEDLARS retrieved a little less than 60% of the total relevant literature within its base. At the same time, on the average, approximately 50% of the articles retrieved were of some value to requestors in relation to the information needs prompting their requests to the system. (268:39)

These figures were averages on all the searches. Swanson has called attention to the high degree of variability in both precision and recall ratios from question to question which in both cases ranged from 0 to 100 percent. He concluded that ". . . the results had little predictive value for the result of any specific subsequent search" (426:133). It should also be emphasized that these results were obtained before online systems were introduced, when machine searching was still in its infancy. It does underline the fact that it is extremely difficult to achieve 100 percent efficiency in all searches. It has also been pointed out that precision and recall seem to have an inverse relationship, that is, that precision is usually increased at the expense of recall. If the search parameters are narrowed, fewer items may be retrieved of which a higher percentage will be relevant, but some desirable items may be missed. Conversely, if looser retrieval criteria are used, recall ratios may be higher, but usually at the expense of precision.

There are some serious methodological and philosophical problems with both these measures which have been recognized and discussed by information scientists. Recall and precision ratios are significant only in comprehensive searches, although it is useful to know the reasons for search failures even for more limited searches. The first problem is that of determining the total number of relevant articles that exist in the database, if you are using essentially the same methods to obtain the measuring instrument and the quantity you are measuring. This difficulty can be overcome to some extent by finding other methods such as asking experts in the field to identify relevant articles or using alternative searching techniques.

A more serious problem is that dealing with "relevance," a problem not only in information science, but in discussions of generation gaps and other issues where profound philosophical differences may exist. Attempts have been made to deal with the problem on logical-philosophic grounds where it appears to be a very illusive concept (392). Relevance is the expression of some kind of relationship. In information systems it is usually based on individual judgments which can vary according to the background and the needs of the individual at that moment. In this sense relevance, like beauty, can be said to exist in the mind or eye of the beholder. Yet there must be some sort of consensus in both areas or we would not have any winners in beauty contests or any working information systems. Some of the general conclusions of studies of relevance in information systems are:

1. The greater the subject knowledge of the judges the higher the agreement on relevance.
2. A corollary to the above is that non-subject specialists, that is, those who have less knowledge on the subject, tend to assign higher relevance ratings than subject specialists. This implies that relevance is not an all or none phenomenon and that some documents may be more and some less relevant than others.
3. Finally, correlations in relevance agreements even among experts tend to vary between 55 and 75 percent.

Saracevic (392:99) codified the definitions of relevance he found in the literature and derived the followng formula: Relevance is the A of a B existing between a C and a D as determined by E where the following terms are used:

A	B	C	D	E
Measure	Correspondence	Document	Query	Person
Degree	Utility	Article	Request	Judge
Dimension	Connection	Text	Information	User

The conclusion reached by another investigator was

> . . . that relevancy is not an inherent characteristic of a document, but rather a function of certain idiosyncratic variables of the user as he evaluates a particular document in terms of his own information needs at that time. (243:45)

The recognition of relevance where it did not exist before is a creative act and one of the methods by which science advances, as Pasteur noted in his statement quoted earlier in which he said that chance favors the prepared mind.

Failures or deficiencies in literature retrieval have been attributed to a number of reasons, some of them related to the database and some to the quality of the search. The ability of the vocabulary to describe the concepts embodied in a search inquiry is a function both of the indexing language and the ability of the requester to verbalize the search requirements. Semantic and syntactic ambiguities of the language are also involved in both instances. A search on MAN *AND* DOG would retrieve articles on dogs biting men and also on men biting dogs, although the incidence of the latter may be small. An article may be indexed with the terms DIABETES *AND* PREGNANCY but may not necessarily include any discussion of the relationship between them.

Some of the failures are attributable to the character of the database, such as the scope and coverage and some of the other factors I discussed earlier. Searchers must also assume some of the responsibility because they should know enough about the database to make appropriate choices. The assistance and guidance given by the system, such as the cross-references in printed indexes and the prompts in online systems, are a factor. Indexing quality and accuracy are also important elements. Studies of indexer consistency have revealed that inconsistency exists to some extent both among indexers and in individual indexers dealing with the same subject matter (296). The question assumes, however, that there is only one way to index a document and fails to consider whether there is any relationship between indexer consistency and retrieval performance (88). Failures attributable to searchers relate to their ability to verbalize a requirement and their ability to search systematically.

Another measure suggested for evaluating literature searches is the "novelty ratio," the percentage of items retrieved which are not already known to the requesters. This is applicable when you are searching systems with overlapping coverages or in checking the results of alternative strategies in the same system. It does not seem fair to use this measure to evaluate a system when a requester has already acquired some of the retrieved references from another source.

Goffman pinpoints two critical issues in the effectiveness of information systems for the individual user, the questions of relevance and quality. Both are surrounded with a great deal of subjectivity and individual specificity, but the latter is even more difficult to deal with than the former. "Almost all large-scale retrieval systems are quantity based in that they treat all literature stored in their vast files as equal in quality" (185:40). Some qualitative judgments are made in all systems in selecting material for indexing, and relevance judgments are integral to the process of indexing, but only one system or technique makes these factors an inherent aspect. In citation indexing both relevance and quality are inherent in the citation process, in that the judgments are made by peers through choices of references which, as Swanson puts it, create "some kind of a relevance bridge" which can be traversed in both directions (426:141).

In the final analysis all information searching is necessarily incomplete or imperfect to some degree. The approach to completeness and perfection is related to the time one can devote to the search and to the ability to develop efficient and effective search strategies. Bernal was very pessimistic about it in his report to the International Conference on Scientific Information in 1958:

> Indeed, the very chaos of present-day publications and the unpredictability of their contents automatically ensures a more or less random sampling of the field of science by the average scientific reader. (34:83)

The situation has changed considerably, I hope, particularly with the introduction of computer-based literature retrieval. Swanson, nonetheless, is probably correct when he refers to literature searching as a "trial and error process." He borrowed the expression, he says, from Karl Popper, the philosopher of science, who used it to describe the way scientific research is conducted where one starts with a hypothesis and gathers information to confirm or to change it.

Chapter 11

Personal Information Files

It is not very often that we are afforded a glimpse into the working habits of scientists, particularly when the glimpse relates to such things as how they maintain their personal information files. Still, it is very likely that all workers develop methods that suit their own predilections and personalities, ranging from the most casual to the most meticulous. Leonardo da Vinci in his lifetime accumulated more than 5 thousand pages of manuscript in which he recorded his ingenious observations and experiments, more or less at random. They were scattered after his death and now reside in various collections in Paris, Milan, Windsor, London, and elsewhere. These notes were not organized until they were brought together and translated by Edward MacCurdy in a collection entitled *The Notebooks of Leonardo da Vinci*, published in 1955. Leonardo himself was aware of the lack of organization of his notes because he prefaced them with an apologia:

> This will be a collection without order, made up of many sheets which I have copied here, hoping afterwards to arrange them in order in their proper places according to the subject of which they treat; and I believe that before I am at the end of this I shall have to repeat the same thing several times; and therefore, O reader, blame me not, because the subjects are many, and the memory cannot retain them. (285:57)

These notebooks do not contain citations from the existing literature of the time but are the product of Leonardo's own fertile mind, as he makes clear himself:

> If indeed I have no power to quote from authors as they have, it is a far bigger and more worthy thing to read by the light of experience, which is the instructor of their masters. (285:57)

Leonardo's failure to organize his notes may be one of the reasons his wide-ranging and innovative observations in art, anatomy, physics, mechanics, etc., were not published until centuries after they were recorded.

Leonardo's notebooks resemble what in past centuries have been called "commonplace books" in which students and scholars recorded notes on their reading, lectures, and observations. Commonplace books of this kind exist in profusion at the Universities of Cambridge and Oxford and elsewhere (90:32). They were frequently compiled by medical students as a means of organizing their lectures and reading. John Locke as a student and Fellow at Oxford in the 1650s and 1660s kept a series of commonplace books on scientific and medical subjects which he arranged according to his own system of alphabetical indexing (142:250). Although we do not seem to have any accounts of his methods of recording and organizing information, the great Albrecht von Haller must have been very systematic to have produced the prodigious number of compendia and bibliographies. From his own accounts we know the value that William Osler placed on notetaking as an essential element in education (347). He early formed the habit of writing abstracts of what he read and adding his own comments, a practice which stood him in good stead when he undertook the compilation of his famous *Principles and Practice of Medicine* (346) which went through so many editions.

We know somewhat more about the working habits of Charles Darwin, thanks to the reminiscences of his son, and the few related passages in his own autobiography. Here he describes how he kept thirty to forty large portfolios in cabinets with labeled shelves "into which I can at once put a detached reference or a memorandum" (103:80). He does not provide us with enough detail to completely reconstruct his literature filing methods, but he does speak of making "short indexes" and "a general and classified index." He claims that "by taking the one or more proper portfolios I have all the information collected during my life ready for use" (103:80). His son Francis Darwin describes how his father kept a separate shelf for the books he had not yet read. As he read them he marked passages by pencil lines down the side, often adding brief comments. He listed the marked pages at the end of each volume and abstracted them onto sheets of paper under various subjects which he then added to the appropriate place in his portfolios (103:128).

The range of options for maintaining personal information files has never been wider than it is today. Some practitioners and researchers may keep no personal information files of any kind. There are always the legendary senior clinicians or research workers who amaze the participants on ward rounds or in seminars, by reciting from memory author, title, journal, year, and page for an article on some controversial issue that arises. It has been suggested that in some cases these situations are carefully rehearsed, but there is ample evidence that phenomenal individuals of this type do indeed exist. Others may avoid maintaining personal files by relying on shared, centralized, or ready-made information files like libraries and indexing and abstracting services.

Learning how to use these public and shared information files is, as I have suggested, an important part of being an effective practitioner or research worker.

I have not said very much about the use of libraries as information storage and retrieval mechanisms, although much of what I have discussed is relevant to their use. I have also assumed that most readers are experienced users of this important information resource. It is important, however, to be aware that each library like each laboratory, despite all the standardization which exists, is a unique and different place, and that each new user must take pains to find out what its resources are and how they are organized. Personal files are by nature highly selective and limited. It is obvious, as Fawcett states, that "No personal filing system can comprise all the relevant material being published in a particular field" (133:84), and that you must constantly return to centralized and shared resources like libraries and indexing and abstracting services to update them.

Stibic insists that: "Every professional person (scientist, researcher, technician, manager, etc.) owns his own personal collection of documents (books, reports, journals, cuttings, photocopies, slides, microfiches, letters, drafts, notes, etc.) that must be well organized and accessible at any time" (421:v). This is probably an overstatement. There are the legendary individuals we referred to earlier who seem to keep all the information they need in their heads. It is also true that there does not seem to be any direct relationship between the effort spent on personal information files and achievement. It has been observed that those with the most carefully organized and indexed files are not always the most innovative or productive. There are, however, sufficient indications in the literature that maintaining personal information files is a recurring and persistent problem. Articles frequently occur in the literature of the life sciences with such piquant and intriguing titles as "Filed and Found" (24), "Keeping Up to Date" (280), and "Where Did I See That Article?" (410). The corollary to the observation that obsessive file keepers are not always the most productive is not that those who keep minimal files are, but that we do not seem to have any evidence of a correlation between the two factors.

Characteristics and Principles

You can usually answer such questions as "Where did I see that article?" by referring to standard information sources like *Index Medicus,* but you can also save much time by retaining answers to recurring questions of this kind in a personal information file. The great advantage of a personal file is that it is "personal," that is, that it can reflect the personal needs and viewpoints of the individual who maintains it. Methods and styles can be developed which are most suitable for the resources in time, staff, funding, and energy available for the task. Personal files save time because they are more accessible, not only by being physically closer to your office or laboratory bench and easier to locate information in than large, generalized files, but also by being intellectually closer because the indexing vocabulary reflects more closely the way the concepts are tagged in your mind.

Personal files can become a by-product of the continual process of keeping up with the literature. Someone who makes a practice of scanning the weekly issues of *Current Contents,* for example, can tag references to incorporate into a personal file or to send for copies for a reprint file. If you are participating in a selective dissemination of information service such as I discussed earlier, items selected from this source can be treated in the same way. This is also true for those who make a practice of scanning specific journals. This points up another great advantage of a personal information file in that it can contain only those materials which have been personally screened and evaluated.

It is not necessary, as with shared systems, to make the scope and organizing principles for personal files explicit, although it is a good idea to be clear in your own mind about these aspects. Personal files may have a great deal more flexibility because they do not need to conform to any standards. However, they do need to maintain a logic and consistency of their own, even if they are known only to a single user. General indexing practices such as clearly defining the terms that are used and the rules for using them should be followed.

Even though each personal file can be unique in its organization and subject matter, examinations of successful models can help to select features which are useful to emulate. You could easily select one of the standard indexes or abstracting media such as *Biological Abstracts* or *Chemical Abstracts* as a prototype and use a similar classification and coding system. You should not, however, usually incorporate material that is readily available elsewhere in handbooks, textbooks, or special bibliographies. It is also a good idea early in the process to establish criteria for removing material from the file as well as adding to it. A primary decision is about what type of material the file will contain: reprints or photocopies of journal articles, research, lecture, or other notes, photographs or slides, or references to other sources. Some materials, slides, specimens, etc., because of their physical nature, may need to be filed separately, but they can be arranged in the same way as the basic file.

There are two primary elements in any information file, no matter what devices or methods, manual or machine, are used to implement it: *physical access,* an ordered location from which items can be retrieved, and *intellectual access,* that is, pointers with cognitive content that indicate where the related items are located. I have been calling these pointers subject headings, keywords, tags, classification codes, etc., but they can also include authors' names and dates. It is possible for a single file to provide both physical and intellectual access when it is arranged in some meaningful way, for example, alphabetically and chronologically by author. A reference, then, to "G. Brown, 1969" will lead you directly to the article in question. An author arrangement may have subject significance also, particularly when certain authors are associated with specific topics. Alternatively, one could arrange the physical file by subject terms in either alphabetical or classified order.

Physical locations which serve cognitive purposes, however, are severely limited because a physical object can reside in only one place at a time. To provide an information item with two locations with cognitive significance, one would have to duplicate the item in the form of another copy of the reprint, lecture notes, etc. An alternative is to file a reference or note in the added location to show where the physical item is located, for example, under another subject in a subject file, or from a secondary author to a primary author in an author file. The problem is also solved by providing auxiliary files on cards or other forms which indicate the physical location of the item. Subject and author files can thus be maintained which point to items in other subject or author locations. The physical file can also be arranged in sequential numerical order as the items are added. The number then becomes the pointer when it is assigned to a subject or author index. These methods are, of course, familiar to us from our use of information sources which embody these principles, such as indexes and abstracts.

The principles regarding the selection of keywords or subject headings are similar in shared and personal files. Personal files, however, are usually more limited in size and subject matter. They also provide the important option of selecting only those keywords which reflect more precisely one's personal interests. Indexers for shared systems are always counseled to index an information item under the most specific applicable term. These admonitions are not always religiously applied because even in large systems such as *Index Medicus,* as we have learned, an item related to a specific concept may be filed under a more generic term if there are only a few items relating to the specific concept in the file. Designers of personal files can almost make up their own rules in this regard. For example, in a file related primarily to nutrition, nutritional aspects may be assigned as specific terms as possible, and subcategories of diseases may be assigned to more generic terms.

Vocabulary control in personal information files has all the options and problems of shared files, no matter what systems or devices are selected. If subject controls are to be maintained, basic decisions need to be made whether one will use a controlled vocabulary or some variation of a natural language approach. In either case you can create an indexing vocabulary as items are added to the file. An alternative is to adopt or adapt some existing subject heading list such as MeSH. Use of natural languages (title words) makes indexing easier to delegate but, as we have learned, tends to scatter references on a particular subject. Chapter headings in textbooks have also been suggested as a source of subject terms. They then become broad generic classes to which the subject entries in the book index can provide a basis for assigning specific subjects. Other sources which have been used are standard nomenclatures and classifications such as the *International Classification of Diseases* (320) and the *National Library of Medicine Classification* (323).

General advice for the selection of subject terms for a personal file is given by De Alarcon (110):

1. Add terms which do not require you to change those selected earlier.
2. Choose terms which most closely reflect your personal needs.
3. Terms may consist of more than a single word, but they should reflect a single concept.

4. Inverted forms should be used when the second word is the important part of the concept, e.g., "Hepatitis, Infectious" rather than "Infectious Hepatitis."
5. Avoid synonyms, abbreviations, and adjectives where possible.
6. Record the meaning assigned to words which may have more than one meaning.

Physical Arrangements

The basic physical file (reprints, notes, etc.) can be arranged in a number of ways. An *author* arrangement has the advantage that you can readily determine whether an article by a particular author is in the file. As I indicated earlier, authors are frequently associated with particular subjects, and in this way an author file can also provide subject access. They can also be filed in chronological order under each author and thus provide an aid in location and in weeding the file. One disadvantage is that access is provided only to the senior author. It may also be difficult to locate anonymous items or items derived from such things as lecture notes.

The basic file can also be arranged *alphabetically* by selected *subject* terms on any specific or generic level. This provides direct access to the material by subject at a single place in the file. The disadvantages are the need to maintain some kind of vocabulary control, the fact that many items cover more than a single subject, and the tendency of alphabetical arrangements to scatter material on related subjects. It is also difficult to determine whether an item by a particular author is in the file.

Subject files can also be arranged in *classified* order with specific topics filed as subtopics under broader headings. If, for example, one chose the categorical tables in MeSH, articles on *Conjunctivitis* and *Corneal diseases* could be filed under *Eye diseases* if they were of minor importance or occurred infrequently in the file; if *Hemic and lymphatic diseases* were an area of primary interest, a second or third level of specificity such as *Anemia* and *Blood group incompatibility* could be included (322). This would bring related material together but also would pose the problem of fitting some specific subjects under a single generic class when they relate to more than one.

A fourth method of arranging personal files is in sequential *numerical* order as the items are added. This provides a discrete and readily interpreted location which is easy to file and to find. Numerical arrangements are essential to some systems like "Uniterm" and optical coincidence systems, as we shall see later. The disadvantages are obvious since numbers have little significance in themselves except perhaps for providing some indication of when the item was added to the file.

The remedy for the disadvantage of the single mode of access files described above is by means of auxiliary files, e.g., supplementary author files for subject arrangements, supplementary subject files for author arrangements, etc. These files can be maintained in various forms, such as notebooks or cards, or with the aid of devices such as the computer. This is accomplished by posting the location of the item in the basic file in the appropriate place or places in the auxiliary files. Thus posting "Brown, JB (1982)" on the subject card *Anemia* will lead you directly to Brown's article on that subject in the author file. Conversely, posting the same term on an author card for Brown will direct you to the same article in the subject file. Numerical files, since they have no intrinsic significance, will require auxiliary files for both author and subject. They are easier to post, however. In any case one may require auxiliary author and subject files for secondary authors and added subject designations for any of the basic arrangements. Auxiliary files can also include references to items in the literature which are not represented in the basic file.

Many of the personal information files described in the literature consist of a single basic file arranged in alphabetical order by some kind of a subject arrangement, or in numbered order based on a classification system devised by the file organizer or derived from an existing classification system such as those used by the National Library of Medicine (323) or in the *Standard Nomenclature of Disease* (320), or based on the chapter numbers used to designate broad subject fields in a textbook. Creager found (94), as a result of a questionnaire he sent out to 381 directors of family practice programs, that systems used to organize reprint and information files were almost equally divided between alphabetical subject arrangements and numerical arrangements representing subject classification.

Schutt (395) describes a system devised for a teaching and reference file developed for a family practice residency program. It can, however, be applied to almost any subject field. In this system the National Library of Medicine subject classification is used. Documents are coded with as many class numbers as necessary to identify the article. In the example given, an article on "Phototherapy in ABO Hemolytic Disease of the Newborn" is assigned the following codes:

WS 300   Pediatrics—Hemic and Lymphatic System
WS 420   Pediatrics—Newborn Infant
WH 420   Hemic and Lymphatic System—Blood

The complete article is filed under the first class number, and photocopies of the first page which contains all the class numbers are filed under the other classes to serve as references to the complete article. The system has the advantage that you need only transcribe a simple alphanumeric code instead of a complete subject heading. A copy of the classification system must be maintained with the system as a means of indexing and accessing the file, but it also has the advantage that it is a standard coding system used in most medical libraries.

Coding systems such as the *International Classification of Health Problems in Primary Care* have also been used as a basis for filing systems (411). Other research workers and practitioners have devised their own classifications for their fields, such as McMahon's (288) "Emergency Medical Filing System," in which the subject is divided into thirty-nine major categories, each with two to twenty-five or more subcategories. There are also other classifications in biology, and medicine and in special fields which can be used for this purpose. The classification systems of some of the standard indexing and abstracting services such as *Biological Abstracts, Chemical Abstracts, Excerpta Medica,* and *Index Medicus* can be used as models or sources from which such coding systems can be derived.

Filing systems which are based on standard textbooks are similar to those which use classification, except that they reflect a different kind of organizational logic. A method described by Fuller (144) seems to have been widely copied, at least as reflected in reports in the literature (410, 444). It is based on a system developed by Maxwell M. Wintrobe, who had a file of over 45,000 articles which formed the basis for his *Clinical Hematology* (468) and which served his other teaching and research needs. Fuller used a medical textbook for the major divisions of an indexing system he devised himself which extended to over 500 subdivisions. A simplified modification of this system which requires no supplementary index and which can be adapted to any textbook was described in 1973 (146). Each chapter provides the file number for the system, and the location of an item is determined by checking the book index to see where its subject matter has been assigned.

Other systems based on alphabetical lists of subject headings have also been described. The use of existing lists like MeSH has the advantage that the vocabulary represents a standard vocabulary that is used in accessing the literature in major shared systems. In personal systems you need not, of course, confine yourself to terms in the standard vocabulary, but can add terms or modify them as appropriate to your personal interests. An example is provided by Gutheil (199), who uses MeSH as a source, but has pragmatically added terms which reflect his own needs. For example, under the term *Humanities* he files all articles on art and literature which are relevant but which do not fall under any other subject on his list. He admits his method may be abhorrent to some, since it is based on what he calls "shredding" his journals as he receives them. He removes all the articles in which he has any interest and discards the rest, photocopying those pages which are common to two articles. On the basis of title, abstract, or summary he assigns a "topic heading" and writes it on the article. For those articles which require two or more topic headings, he writes them on separate sheets of paper indicating the main heading under which the article is filed.

It becomes obvious that, although general principles can be identified, there are almost as many different individual filing systems as there are individual files. As graduate students, individuals are encouraged to develop citation files on cards, with complete citations and abstracts on which they compile everything relevant to their research interests. Some are able to retain these exemplary habits throughout their professional lives, but most find it necessary to develop less formal methods of various kinds that frequently embody several methods either concurrently or successively. One investigator (personal communication) described his system as "not ideal" but as one which "worked" for him, because it could be accommodated to the amount of time and energy he was able to devote to it. He keeps only the last 10 years of the journals he receives personally, discarding them after this period except for the "classical" articles which he either clips or photocopies for his reprint file. He also scans selected journals on a regular basis in his institutional library, listing those articles which seem to be of potential interest in a notebook. On the basis of these notes he decides which he will want to read when he has more time. After reading them he decides whether he wants to retain them in his personal file in the form of reprints or photocopies. For articles in his personal journals he writes the citations on separate sheets of paper for those articles in journals he retains intact and clips the others. Other sources of articles are a Selective Dissemination of Information service from his institutional library to which he subscribes and abstract journals which he receives from specialty groups. He files the photocopies, reprints, etc., in three separate subject files. One is a "current file" which relates to work in progress on which he is in the process of preparing a report. Another is an "active file" which deals with those research topics with which he is currently involved. Finally, there is a "reserve file" which includes articles not related to his current activities but which he regards as having general or potential interest. He uses the same list of subject headings, which he developed himself, for all three files, but subdivides some of the subjects in his "active file" by the name of the author or research group which has been responsible for a corpus of work in that field. He admits that this is not a method that would work for everyone. He is also open to suggestions to improve his system, providing they are feasible in terms of the kind of time and staff support he has available. He is aware of and hopeful about the potential of the computer in these kinds of activities but has not yet been able to find the computer-assisted method which fills his needs.

Mechanical Aids

The computer is the latest in a series of mechanical devices and methods which have been proposed and used in personal information systems. Not all personal information systems are amenable to these aids. Some systems like citation indexes are particularly designed for mechanical assistance. Even citation indexes, however, can be managed manually if one wishes to take the time and trouble. Each cited reference can be written or typed on a separate card or sheet of paper, and each time it is encountered in a source article, the source article could be listed on the card. This may prove a useful technique to someone who wants to maintain this kind of a record, but in a sense a manual application of this technique defeats the purpose since the citation is a method of finding the source article and not vice versa.

Manual systems have their own advantages. They have little "hardware" cost unless you include the rising cost of filing cabinets. They tend to have little "down time" unless the lights go out and you are left completely in the dark. If properly planned, they can be easily expanded if you have the necessary space.

They usually have low maintenance costs unless you count the personal and staff time involved in maintaining them, in which case they may be quite costly. It is this last factor which makes some machine methods attractive and which has stimulated the application of new technology to the problem.

Even notebooks and cards can be considered forms of mechanical aids. Notebooks to some extent may seem like the commonplace books of our ancestors, but they can be modified by using looseleaf forms and by coding and indexing entries so that they can be easily found. A recent author (376:100) describes a looseleaf notebook system in which a page is reserved for each subject in alphabetical order. Citations and brief abstracts of articles are entered on the appropriate subject pages.

Other methods can be adapted to notebooks or cards. A number of these are described by Jahoda (233) and Stibic (421). One is the "uniterm system" to which I referred earlier. Hellmers (212) uses notebooks for a file based on the "Modified coordinate index system" developed in the 1960s for the Engineers' Joint Council. The reprints in his file are given consecutive numbers as they are added. The numbers are posted on the appropriate subject pages in the notebook. By matching reprint numbers on the pages for those subjects he wishes to coordinate, he is able to identify the pertinent reprints.

The system is more readily adapted to formatted cards which can be pre-printed. The cards have ten columns and the document numbers are posted in the appropriate columns according to their terminal numbers; all numbers ending in 0 (30, 270, 350, etc.) are entered in the first columns, those ending with 1 in the second and so on. By pulling cards for two or more subjects and matching the columns you can quickly identify those documents which have the desired characteristics. Broadhurst (60) describes such a system for a file of 6,000 reprints in psychology. Pearson (356) describes another such system in medicine as an index to reprints, photographs, and slides. Each file is prefixed with a letter, R for reprints, etc., to indicate its location. He also uses journal titles, year of publication, and author as keywords so that they can also be coordinated.

Another device which was widely used during the 1950s and 1960s involves printed cards of various sizes with holes punched around the outer edge. The holes can be notched to represent coded subjects or numbers and then sorted by needles to retrieve those cards with the desired attributes. Each document is represented by a single card on which the full citation and abstract can be recorded. The cards can be coded either "directly" or "indirectly." In direct coding a single meaning is assigned to each individual hole. A 5 by 8 inch card has ninety-seven holes around the edge and thus can work well for any system with ninety-seven or fewer subjects or aspects to control (233:63). A needle inserted through a hole which is coded to represent a particular subject will cause those cards which had been notched at that position to fall away. There are many modifications of these methods. On directly coded cards needles can be inserted through two or more holes to produce coordinated searches. The cards can be stored in random order and returned anywhere to the file, although in some systems they are filed alphabetically by author. The holes can also be indirectly coded by using combinations of holes in assigned fields used to represent numbers. Four holes in a single field can represent any number from 0 to 9. A combination of two fields expands the possibilities to 99, through 999, and so on. There are a number of other modifications, some of them involving edge-punched cards which can be either shallow or deep punched to increase the coding possibilities.

One system described in 1961 used an edge-punched 3 by 5 inch card "because it readily fits the pocket" (403:613). It had fifty-one peripheral holes and used a combination of both direct and indirect coding. The first eight holes were indirectly coded for authors' names following a system developed by *Chemical Abstracts* which breaks names up into 100 alphabetical categories. The next thirteen holes were directly coded for general categories in which each hole represented such concepts as *Morphology, Metabolism, Genetics,* etc. The other holes were assigned in groups of eight (ninety-nine in each category) which could be indirectly coded for subjects, species names, etc.

A single card in an edged-notched system can provide access in a number of ways. If they are used as an auxiliary to a reprint file, the cards also need to include pointers to that file in the form of numbers in a numerical file or authors to provide access to a reprint file arranged by author. A disadvantage of edge-notched cards is that only a limited number can be handled at any one time by a sorting needle, and it may take some time to sort through a large file. Complaints have also been heard that cards that have been punched to drop in a particular sort may adhere to adjacent cards. Mechanical and electrical devices have been developed to overcome these shortcomings, but some systems of this kind initiated with high hopes have been abandoned.

Another system which predates the computer and which also had a considerable vogue, though less in personal information files than in larger institutional files, involves the use of what are variably called "punched feature cards," "optical coincidence cards," or even with some suggestion of coyness "peek-a-boo cards." They operate on the principle that a hole punched or drilled in a specified location can be made to designate a particular document. The cards, which are sometimes made of plastic, can be of various sizes and ruled off in grids with each square representing a unique number. Thus a 6 by 10 inch card can be designed to hold as many as 1,000 numbers, and cards up to 40,000 capacity have been used (233:57). The method essentially is an application of the uniterm system. Each card is assigned a subject or some other designation, and the square on the card which corresponds to the number assigned to the document is punched or drilled so that light will shine through at that point. For subjects with a single facet the card assigned to the subject will indicate all the related document numbers in the file. Where two or more subjects must be coordinated, the cards are placed one over the other, and those points at which the light shines through will designate those documents which are common to these cards. Large commercial systems have been developed which use semiautomatic precision drills and light display sources.

One such system was set up to access a large file of necropsy reports on zoo animals. A card with 100 by 100 coordinates was used representing 10,000 numbered locations on which reports could be posted. Two classified vocabularies based on a list of animal species and a list of diagnostic categories and procedures were used (196). A smaller "homemade" system using a 5 by 8 inch card, with coordinates 16 by 30 providing 480 locations, was devised for patient records in a research project. Each patient record was assigned a sequential number and punched on cards which characterized the case by age, sex, diagnosis, anatomic location, type of therapy, and result. By selecting those cards for age group 30 to 40, male sex, bursitis, shoulder girdle, ultrasound therapy, and symptom-free results, and superimposing them, the author could easily determine "how many young males with bursitis of the shoulder had been clinically cured following ultrasound therapy" (400:420). Systems have even been devised using the standard Hollerith card, sometimes called the "IBM card," which we are admonished not to fold, spindle, or bend. The card is preprinted with 800 positions which can be punched with a hand punch. When superimposed over a light source they can be used like any other optical coincidence card (233:58).

Computers

The phenomenal growth of computer technology in the last few decades, particularly with the personal or microcomputer, has opened up a great many more options for the management of personal information files. Computers promise greater speed of access, automatic transfer of data in electronic form from external databases, and other benefits. The technology is changing so rapidly that comments on the capability of any particular system may go quickly out of date. There are, however, some principles and observations which deserve consideration before selecting a specific computer-aided personal information system.

Many different programs of instructions have been written to enable computers to store, process, retrieve, and print data. They have been prepared for computers of different size and manufacture, ranging from the large institutional or "mainframe" computer to the smaller "minicomputer," which is rapidly approaching the mainframe computer in its capability, to the new class of "microcomputer" or personal computer, which is becoming more and more accessible in cost and ease of operation. At this point in development there is a great deal of incompatibility among systems in "hardware," the physical components, and in the "software," the instructions written by programmers. The programs, moreover, are written in various "languages" like COBOL or BASIC which provide the codes which instruct the computer to accomplish specific tasks.

It is possible, of course, with some expenditure of time, energy, and endurance to write your own program to manage a personal information file. The usual advice, however, is to try to find an existing program which will accomplish most of your objectives, even if you have to make some compromises. It has been estimated that it takes between 300 and 500 hours to learn how to program adequately in one of the relatively simpler languages like BASIC. It can also be costly to hire someone to write a program for a special application, because according to industry standards only about twenty instructions can be written in a day. A typical program of 600 instructions at $30 an hour, therefore, could cost as much as $7200 (21). There are also something like 200 different computer languages currently in use which cannot be "understood" by a computer unless it has been programmed with an appropriate "operating system." Programs, therefore, must be chosen with a view toward the particular kind of computer—mainframe, minicomputer, or microcomputer—on which it will run.

Large numbers of programs exist for all these classes of computers, some with special and some with more generalized functions. Among the more general programs are the "database management systems" which enable you to set up the form in which you wish to maintain your records, to add, change, or delete any of the records, and to access them individually or sort them for various kinds of listings. These systems can be used for maintaining inventories and fiscal records, but can also be adapted for personal information files. They are not all the same. Some permit you to use Boolean operators to sort and print records. Frequently, they need to be "interfaced" with other programs to be able to edit or format reports. Among these are "text-processing" or "word-processing" programs which are also used for correspondence, preparing papers for publication, and other purposes.

It is possible in some instances to use word-processing programs alone as a basis for personal information files, since they usually provide the capability of accessing any particular word or string of characters in the text. They do not usually have any means of designating "fields" such as author, title, etc. They are slow, therefore, because they must search each word of a file sequentially to find the designated records.

Fields in computer records can be either "fixed" or "variable." Fields for author, title, etc., are fixed if a maximum number of characters are allocated to fill them. For example if 100 characters were allocated for the title field, titles longer than 100 characters would have to be shortened, and titles shorter than 100 characters would still occupy the same amount of space in the record. On the other hand, no limits are set on the number of characters in a "variable" field which occupy the same amount of space as the number of characters needed for that record field. Each method has its advantages and disadvantages. Fixed field records are quicker to use in random access searches, but short records occupy as much space as long records, and therefore larger storage capacity is required. This can be a significant factor when storage available is limited, as with microcomputers, or when you must pay for storage in mainframe systems on the basis of the amount of space occupied. It is also more difficult for fixed field systems to interface with online systems, which usually require variable field formats. Some systems are mixed in that they provide variable lengths for some fields, such as author and title, and fixed length for such fields as publication date and record number.

Computers have been used as an aid in managing personal information files since the early days of computers (measured in terms of a single decade) when they were used in the batch mode for such things as producing multiple cards for filing systems (66). Since then, many programs have been written for institutional mainframe systems, with such names as *Access, Index, Scrapbook,* etc. They are usually written for a particular institution's configuration of hardware and operating software and are not easily translatable to another institution's computer system. Most large computer facilities, however, have commercially developed database management systems which can be adapted for personal information files. A system of this kind developed by the National Research Council of Canada which operates on a large time-sharing computer was recently described. It uses a resident database management system in conjunction with a text editing system which enables users to sort and print their files by author and subject (278).

Another form of access to mainframe systems is provided by some of the same vendors who bring you online access to *Index Medicus, Chemical Abstracts, Science Citation Index,* and *Biological Abstracts.* They can accommodate personal files of any size, and provide for entry, access, and printing of data both offline and online. Charges vary from one vendor to another, but they can all be rather costly. These services provide users with their own passwords and use of all the sophisticated searching and reporting features they offer.

These are obviously large systems and not suitable for everyone. Microcomputers have recently brought computer-assisted management of personal information files within easy reach of most individuals. There is a wide variety of programs for microcomputers currently available for this purpose, although they are still not as easy to use as one might wish. Personal computers have still not reached the point where they can be used like electrical appliances simply by plugging them in. It still takes hours to learn how to operate one, and each new applications program must be learned separately along with the different meaning of the control characters in each program. Their usefulness depends on the ingenuity and resourcefulness of those who have written the programs. They all present varying capabilities as to how keywords are entered, how records are entered, whether words can be truncated, how they are searched, etc. They are much more limited than mainframe systems in terms of storage capacity, both in the computer and in auxiliary storage devices. Microcomputers range in memory from 16K or 16,000 characters or less to 64K or 64,000 characters, although larger memories are becoming available. At an average of 200 characters per record, this means that a microcomputer of 64K can hold an average of 300 records before they must be stored externally on disks or other storage media.

Disks for microcomputers also are limited. The standard 5½-inch "floppy disk" holds 128,000 characters which can average 600 records. In some programs this may require using journal and subject codes instead of spelling them out. This situation is changing rapidly, and floppy disks of 500K and larger are being spoken of. There are also "hard disks" for microcomputers currently available which, although more costly, can hold 6 million characters or more.

Programs for microcomputers which perform some of the same tasks as programs prepared for larger computers are proliferating at an ever increasing rate. Among these are database management programs which can be used for personal information files. They range in cost from $15 to $1,000, whereas the costs of such programs for mainframe systems can go up to $200,000 or more (430). They vary greatly in other ways as well, such as in kinds of "disk operating systems," record length, memory required, and the things they are able to accomplish. Most of them have been written for business use such as maintaining inventories, fiscal records, and customer lists, but they can be adapted for personal information files. A characteristic of most of them is that they use fixed field formats, creating some of the problems discussed earlier. The problems of transferring data to and from programs, can be overcome to some extent by using them in conjunction with text- or word-processing programs for creating, deleting, and editing the file.

Programs which are more specifically designed for bibliographic data are also becoming available. Some of these, like the Institute for Scientific Information's SCI-MATE, can be adapted to run on different makes and models of microcomputers. It is also more "user friendly" in that the program prompts you by "menus" or lists of options from which you can choose, to enable you to create, edit, sort, and print records. SCI-MATE consists of two components. The "Personal Data Manager" provides the basis on which to create and maintain a personal information file, as well as to accomplish other tasks available in database management systems. The "Universal Online Searcher" provides you with a single series of commands which enables you to search many of the major databases such as MEDLINE, SCISEARCH, CASEARCH, and others with the same protocols. These components are interfaced in the system so that records retrieved from online databases can be transferred to personal files under control of the "Personal Data Manager." The maximum length of a single record is 1,894 characters, but provision is made for continuing on additional records. Users can also create their own formats for different kinds of files.

This is, of course, not a very complete description of either the capabilities or the limitations of computers for personal information files. One problem, although it may also be regarded as an asset, is the wide diversity of choices. A recent survey of portable computers, for example, listed some seventy-five different machines ranging in price from under $100 to almost $10,000, in weight from 4.2 ounces to 36 pounds, and in memory from 2K to 704K (470). They represent a rapidly changing technology in which developments occur everyday in both hardware and software. They add a whole new range of options which need to be carefully considered by those who are setting up personal information files.

They are not necessarily the best answer in all cases. It is sometimes much better to adopt a simple manual system which will be maintained even if it does not provide all the answers, no matter how seductive the new technology may be. Violin teachers have an expression which may have some application here. Beginners on the instrument start to learn their fingering by playing in the "first position" on the fingerboard. As they progress they move up so that the same pitches can also be played higher up on the keyboard on lower strings. This is useful sometimes when one wishes to change the timbre of a note for expressive reasons, or to make it easier to execute a particular passage. At other times, however, it may be stimulated by a desire to show off one's virtuosity even if it makes the fingering more difficult. At times like this violin teachers have been known to exclaim: "It's no disgrace to play in the first position."

Chapter 12

Scientific Communication: Past, Present, and Future

The growth in electronic technology in recent decades has been so rapid that it is hazardous to predict what its impact will be on the dissemination, storage, and retrieval of scientific information in the future. It provides great opportunities for the marvel mongers who seem to have a wonderlust for extrapolating from every technical innovation to a vastly changed social order. There do not seem to be any aphorisms about the future as meaningful as Santayana's "Those who cannot remember the past are condemned to repeat it," but being aware of the fate of some of the predictions of revolutionary change in the past should make us cautious about predicting the future. The Sunday supplements after World War II were full of predictions about how helicopters were going to transform our transportation systems and how every home would have a landing pad on its front lawn. They failed to give much consideration to how travel in three dimensions would compound the traffic problems they were having in two dimensions. An example of technical change which has had great social impact is the automobile, yet it is doubtful whether anyone 10 or 20 years after it began to appear on our city streets had any idea about what form the changes in our cities and in our life styles would have.

We are living in a highly transitional period in scientific communication when older systems and technologies are being severely tried and questioned, when newer technologies are becoming available and new modalities are being tested. They must all be examined in the context of the past, when the systems we are using were developed in response to social and intellectual needs which may not have changed, and in relation to the present, when we must make our decisions about how we are to deal with our information dissemination problems today.

We have been presented with a revolutionary new technology in the computer which has already had a great impact on our information services. Many of the printed services we use would not be possible without the computer. This new technology makes possible the elimination of repetitive typing in revising

174

and transmitting texts. Texts stored in electronic form can be easily changed, reformatted, and transmitted over long distances without being replicated. Electronic transmission of texts from authors to editors and reviewers is already being used, with a consequent reduction in time lag. The means for book printing on demand are already available. Texts for specialized and limited editions can be stored electronically and produced in printed copies only when required. This may help make it easier to "publish" these kinds of works than when the market for them has to be predicted and the books have to be stored in warehouses in anticipation of demand. The question still remains whether any of these innovations will radically change the ways we organize, synthesize, and use information. The automobile may have greatly changed the speed with which we get from one place to another, but it has not to any considerable degree changed the purposes of our trips or what we do when we get there.

New Technology

The rate at which changes have taken place in the development of the computer is exemplified in an article on "personal computers" written in 1982:

> If the aircraft industry had advanced as spectacularly as the computer industry over the past 25 years, a Boeing 767 would cost $500 today, and it would circle the globe in 20 minutes on five gallons of fuel. (438:87)

The increase in computer storage capacity has averaged 35 percent a year for the 20-year period 1960 to 1980 (56). In 1980, a computer microprocessor which weighed less than a pound could hold eight times as much information, was twenty times as fast, and used 56,000 times less energy than the first electronic computer built in 1950, which weighed 30 tons (269:16).

Some of the futuristic predictions on the size and speed of computers and on the cost of energy required by them are dazzling. Electronic storage on random access chips which make up the computer's memory has increased phenomenally. In 1973, one chip could store 256 bytes (each equivalent to one alphanumeric character). By 1982, a chip of the same size could store 64,000 bytes, and it was expected that capacities of 1 million bytes would become available "in a year or so" (145:93). This can be translated into familiar terms when it is estimated an average book of 300 pages and 1,500 characters a page requires a storage capacity of 450,000 bytes (the group of binary digits or bits that represent a single character in the computer). A computer company executive estimated that it would require 70 trillion bits to store the 20 million or so books in the Library of Congress. Using figures he supplies for current computer costs, only $120,000 a year would be required to make a 100,000 volume library available instantaneously to any user anywhere who had the machines to access it. While this may seem like a large sum for a single institution, it may be an economical way to supply information to a number of institutions or even individuals (55).

Even more phenomenal advances in computer storage are predicted. Speculations have been offered on the development of "biochips" using viruses and other living organisms, based on the rationale that if the DNA molecule can store so much information in such a small space, then plasma-based storage devices can be developed which can store as many as 800,000 billion characters in a cubic foot of space. Even with current technology the theoretical limits for magnetic storage on a 14-inch disk, with an access time similar to today's, has been estimated to be 2.5 billion characters (145). Using our previous estimate of 450,000 characters for an average book, that would translate into over 5,000 books on a single disk.

Videodiscs and optical disks present another technology in which revolutionary developments are expected in information storage and retrieval. It is sometimes difficult to sort out and evaluate what you read about any of these new media, because it is not always clear whether the authors are reporting current or potential capabilities. Without trying to explain the technical characteristics of these electronic disks, three types which are currently available can be described. The videodisc, which is essentially an "analog medium which can be used to store digital information" (352), exists in two forms: "capacitance" systems which are read with a stylus and optical videodiscs which are read with a laser. Both of these formats require copies to be made from an expensive master, and the costs must be amortized over a number of copies. They therefore can only be "read" and not "written" on. The optical laser disk is a "read-write" medium which can store images in digital form and can be easily copied and reproduced. It is also capable of large-scale storage. One species of optical disk is said to be able to store the contents of the *Encyclopedia Britannica* on a single disk (188).

The great advantage of storage in these media is that illustrations and even motion pictures can be stored along with textual material. Other capabilities include random access, "freeze frame" (the ability to stop on a particular frame in a moving sequence), and control of the speed with which frames are advanced. One problem is that, although the resolution available on standard television receivers may be adequate for motion pictures or illustrations, it is not yet adequate for reading the typical printed page. Research, however, is under way sponsored by the Library of Congress to provide high-resolution screens and laser printers so that the medium can be used to preserve deteriorating paper documents and produce copies on demand. There still seem to be problems of permanence of the new medium (325). Optical disks are regarded as having "a significantly longer lifespan than magnetic media," but "Best estimates are that optical disks will last 10 years or more" (352:411), which is hardly reassuring.

Interesting technical developments and projections are reported in other areas as well. The price and size of satellite communications systems are said to have decreased by a factor of ten between 1970 and 1980. Direct broadcast satellite systems using 2-foot antennas which will cost about $100 each have been projected (106). Most online searching today is done at a 1,200-baud rate, but transmission rates of 9,600 and even 19,200 baud, or thirty characters a second, are being talked about (128). The number of online databases increases every year. In 1983 130 new databases were added, to increase the total to over 1,700 (339). A Delphi study in 1980 predicted that 50 percent of all existing abstracting and indexing media and 25 percent of all existing reference books would be available only in electronic form by the year 2000 (269:61). Another report in the same year predicted that by 1987 half the households in the United States would have "intelligent terminals" (269:16). Although more than half the interval of time has transpired since the report was issued, it seems highly unlikely that we are anywhere near half way toward that goal.

What are the implications of all these projections?

- The technology will become less expensive and therefore accessible to more people.
- Greater and greater amounts of information will become available in electronic form.
- Programs for accessing and manipulating the information will become more powerful and therefore easier for the uninitiated to use them.

These changes, however, will take place only if there is public acceptance of the new technology, and radical changes of this kind sometimes take longer than we anticipate.

Cultural lag is not the only reason that some of these predictions may not be realized. The problem is not solely that computer terminals cannot be taken into some places that are accessible to the book. Visions have been evoked of book-sized electronic devices into which text modules equivalent to a book can be easily inserted and randomly accessed (269:60). Nonetheless, until the technology improves greatly books will still be regarded by most individuals as a faster and more convenient way to read textual materials. One of the current obstacles is the need to interact or "communicate" with computers by means of a keyboard. Individuals, including scholars and scientists, who need to interact continually with computers can develop great keyboard skills, but for many this can provide a serious obstacle. Voice-responsive and voice-producing computers are already available, but the time when we can carry on an intelligent conversation with a computer still seems a long way off.

In many ways, as we shall see, the new technology sometimes raises as many problems as it helps to solve, but this has been true about almost every new technology. One of the dangers is that because of its novelty we will try to depend on the new technology to solve problems that are economic, social, and intellectual rather than technical. Some of the efforts to solve these problems have included the development of new publication formats.

New Formats

One of the basic problems with the scientific journal since the beginning is that it has been forced to serve two different functions, that of a vehicle to deliver information to a specific audience and that of a repository from which information could be retrieved on demand. Once a journal issue has served its role as a vehicle to disseminate information, it becomes primarily a location device, a kind of a shelf marker for journal articles. It is recognized that even specialized journals deliver articles which interest only a minority of their readers. "A scientist subscribing to a journal," says one critic, "is forced to pay for twenty or thirty papers which do not concern him in order to get one he wants" (359:3). The need for specialized journals has grown as a result of the increase in interdisciplinary research. Among the consequences have been greater scattering of the literature, a reduction in the size of the readership for many journals, and increases in cost. The system has also been criticized for its inability to accommodate all the material that deserves publication, for restrictions on the length of articles, and for inordinate delays between the completion of a manuscript and its appearance in print.

Proposals were made to modify the system long before the computer appeared on the scene. One of the most radical proposals was made by J. D. Bernal, first in an appendix to his *The Social Function of Science* in 1939 (32). He elaborated it about 10 years later in a report to the Royal Society Scientific Information Conference in 1948 (33). Bernal proposed the establishment of central clearinghouses or "National Distribution Authorities" which would play an administrative–clerical role in receiving and processing scientific papers under the jurisdiction and review of established scientific bodies. These central agencies (he also called them "Scientific Information Institutes") would print, index, duplicate, and distribute these papers to subscribers in accordance with subject profiles they had established or on demand based on titles identified in the announcement services.

Proposals to distribute scientific reports as separates were in fact not particularly new. As early as 1922, a proposal was made for central depositories from which supplementary materials such as tabular data and explanations and discussions, which might have limited interest, could be supplied on request (10). A more detailed scheme of this kind was presented in England in 1926 (363). Watson Davis in the United States established a "Bibliofilm Service" in 1933 under the American Documentation Institute to distribute in microfilm articles in print in the major federal research libraries in Washington. He also made the facility available for anyone who wished to supply supplementary data which could not be printed in journals (107).

Bernal's proposal met with some support at the time. Some were willing to accept it only as an auxilliary distribution system as long as journals continued to be available (374). It may also have met with some opposition as much for the radical position of his politics as for the radical nature of the proposal. In any case he withdrew it from discussion at the Royal Society Conference because of what he felt were a number of unresolved questions.

The issue was reviewed in 1958 in response to a proposal by UNESCO that "in view of the increasing inadequacy of the scientific periodical as a method of communication, a long-term study of the whole problem should be commissioned" (78). The investigators responding to this charge reviewed proposals and criticisms of various "alternatives to the scientific journal" up to 1960. Their conclusion was that distribution of separates was "impractical, inconvenient, a deterrent to scientific progress, a threat to scientific freedom and prohibitively expensive" (359), although no conclusive evidence seemed to be presented to justify any of these claims.

The National Institutes of Health from 1961 to 1967 conducted an experiment which met some of the criteria of the Bernal and similar proposals. Information Exchange Groups (IEGs) were established in the Institutes by small groups of investigators who found journal publication too slow for their purposes. They sent their unrefereed papers to a central office where they were duplicated and forwarded to those who had been accepted as members of the group. Any bonafide researcher in the specific field could apply for membership. Seven such groups were established between 1961 and 1967. Price and Beaver investigated the first IEG (on oxidative phosphorylation) which had been established in 1961 with 32 members. By 1965, the membership had grown to 592 and the number of papers distributed had doubled every 7 months to close to 400 a year (373).

The membership had risen to 3,625 in all the groups by 1966, and over 140 manuscripts were being distributed each month. It was estimated, by projecting the current rate of growth, that the membership would be up to 14,000 in 2 years and the National Institutes would be distributing as many as 30 million copies a year (4). It could almost be regarded as an example of ontogeny recapitulating phylogeny, or as Ziman commented: " . . . a method of rediscovering the scientific journal." He added, "Given another decade or so, they might even have rediscovered the referee!" (475:320). There were, in fact, complaints that the service was bypassing the standard reviewing functions, that the medium was elitist, that it was redundant since the papers were later published, that it complicated citation practices, and that it failed to promote discussion, which was one of its aims (119). Accordingly, in 1966 it was recommended that the "experiment" be canceled and that the respective responsible societies take over the responsibility of distributing the papers as an adjunct of their journal publication services. At the same time, it was nevertheless noted that the initial acceptance by the scientific community indicated that there was a real need for such a medium (85).

A number of publication formats have been tried in an effort to answer some of the perceived deficiencies of the scientific journal. Many of these are still being used. The following are examples.

1. *Camera-ready journals* which eliminate typesetting by printing the author's article in the typed format submitted.

    Publication lag time can be reduced to 6 weeks or less, although readability is in some cases sacrificed. This objection could be overcome, however, if authors could submit their manuscripts in machine-readable form which could be used for computerized typesetting.

2. *Synoptic or synopsis journals* which publish only summaries, with the full articles made available in microform to accompany the issue or to be supplied on demand.

One such journal, the *Journal of Chemical Research,* has been published in this way since 1977. Rowland surveyed its readers in 1981 and found little enthusiasm for the medium. Some authors feared the full articles would not be read; others regarded the medium as having lower standards than full text journals, although they recognized the advantages of greater speed of publication (387).

3. *Rapid publication journals,* sometimes also called "letter journals," which are either typeset or produced from authors' typescripts.

One such series, published by the International Research Communications System (IRCS), adopted one of Bernal's suggestions by publishing all the reports submitted in microfilm or microfiche in the form of a "Library Compendium." They are also repackaged in one to ten or more of thirty-two journals, depending on their subject relevance. The reports are limited to 500 words, and publication in this format is said not to preclude publication in extended form in another journal, although some editors may reject them as having prior publication. IRCS also publishes a monthly classified list of all the articles in the system, from which individual reports can be ordered on a subscription basis. A method standardizing the citation for a report no matter how many sections it appears in has been tried, but the whole service does not seem to have been generally accepted by all the indexing and abstracting services. There are a number of other rapid publication journals in various subject specialities which are well covered in the secondary media.

4. *Peer commentary journals* which attempt to overcome the criticisms of the referee system by collecting and publishing reviews along with the main article.

Garfield characterizes these as a step in the opposite direction of rapid publication journals because they tend to increase the time lag (154). One of the few examples of the medium is *Behavioral and Brain Sciences,* which publishes about four articles in each issue with brief commentaries from about thirty-five other authors who have contributed to the subject.

There have been other efforts to modify the conventional format in physical forms:

5. *Microform journals* published exclusively in that format are a rather recent innovation.

Microfilm and microfiche have been used for a long time as an auxiliary form of publication to reduce both storage and distribution costs. Many publishers make microform versions of their journals available within 12 months after the publication of the last issue of a particular volume. Prices are apparently fixed by negotiation with the publisher, sometimes at levels equal to the original subscription price and sometimes at reduced levels. Some publishers restrict sales of microforms of current volumes to subscribers. Original publication in microform was introduced in 1959 by the Wildlife Disease Association, which publishes its *Wildlife Diseases* on microcards (236).

6. *Miniprint,* a method in which reduced text is produced by print rather than photography, has also been used as both a supplementary and an alternative form of publication.

Books in small formats can be said to go back to the famous Dutch printing family, the Elzeviers, in the seventeenth century. Miniature books, some of them less than an inch in size, have been produced and collected ever since then more or less as curios. Miniprint has been used by contemporary journals such as the *British Medical Journal* for printing all or parts of their textual materials. Although some miniprint can be read with the unaided eye, generally some kind of magnifying device is required. Rowland asked his respondents for their preferences between miniprint and microfiche as a reading format. They voted thirty-seven to seven in favor of miniprint, largely because it did not require the intercession of a machine (387).

And finally it is also possible to speak of:

7. *Audio and video journals,* series which are issued on a continuing and periodic basis in the form of audiotapes and videotapes, although they are primarily for continuing education.

Electronic Journals

The most recent change in format proposed for the scientific periodical is the "electronic journal," a journal which is produced, stored, and read electronically. It has been estimated that the process of submitting a manuscript, editorial review, checking proof, and distributing the printed copy can take as many as seventeen mailings between author, editor, printer, publisher, and eventually the subscriber (67). Computers can be involved in this cycle in many ways. Authors can produce their manuscripts in machine-readable form with word-processing programs and transmit them on disks or directly over networks to editorial offices. Editors with the aid of computer-based routing programs can select and send them to the most appropriate reviewers in the same way. Manuscripts can be edited, revised, and finally printed with the aid of computer-driven photocomposing equipment or distributed solely in electronic formats. In all these steps it is possible for the computer to minimize the need to retype the copy and thus significantly reduce the lag time in getting the article "published."

Recent studies have shown that as many as half of the scientific and technical articles prepared for journals and more than half of the American doctoral dissertations are now being produced electronically. The Association of American Publishers is in fact hoping to develop industry-wide standards that will make it possible for authors to submit their copy in a form that can be processed electronically on any machine (1).

A great many of the scientific journal articles read every year are, in one way or another, distributed in the form of separates. In addition to browsing, journal readers are using indexes and abstracts and other secondary announcement services such as *Current Contents* as alerting devices. The British Library Lending Division now sends out 2 million photocopies of the articles in its journal files every year (200). The question has therefore been raised whether it is still necessary to continue to distribute, in costly printed formats, collections of articles, many of which will not be read by their recipients, or whether we should rely on providing online access over networks to the same journal articles, many of which already exist in electronic form.

Many advantages could be said to accrue from the latter system. It could introduce us to a world in which a journal volume is never off the shelf or absent from the library, or worse "at the bindery." It may make possible small group publishing. It has been estimated that even a minicomputer could support a journal for 5 to 10 thousand readers at a cost of about $5 for each article accessed and still allow sufficient margins for profit and marketing (441:201). Various enhancements are also possible. Full texts of articles can be made available for online searching. It has been suggested that electronic journals could allow referees and even readers to tag articles with scores that could be made available in the retrieval process (441). They could also free up publication schedules because a contribution could be put online (published) at any time it has cleared the editorial process without waiting for a complete journal package (vehicle) to be assembled. Some data which are unwieldy when stored in journals such as nucleotide sequences of DNA, of which it was predicted that over 1 million would be discovered in 1981, can easily be manipulated and studied when stored electronically (3). It has even been suggested that electronic journals could add capabilities which do not exist with printed journals, such as animation and audio (215).

There are already a large number of full-text journals available online, but they are largely a by-product of their print versions which still provide the primary distribution method and revenue, with the electronic formats available as supplementary and alternative forms of storage and access. Full texts of such newspapers as the *New York Times, Washington Post,* and the *Economist,* as well as many technical and business journals, are available online. In 1981, the American Chemical Society made available online the full text of sixteen of its journals. The *New England Journal of Medicine* was made available in a full-text online version in late 1983. The Elsevier Company, which took over the IRCS series of journals, has made that entire file of articles, as well as twelve other journals it publishes, available online since January 1980. These are not put online until 6 weeks after the print version is published, however, which underlines its use as an auxiliary distribution medium (127). Moreover, the acceptability and cost effectiveness of the medium does not yet seem to have been demonstrated. This is illustrated by the failure of a proposal by six of the largest scientific publishers in a project called ADONIS (Article Delivery over Network Information Service) to store some 1,500 journals for both online access and distribution on optical videodiscs. Three of the publishers dropped out early in 1983, expressing concern about their ability to market the product, and by the end of the year the project was abandoned (437).

Journals available solely in electronic format have also been attempted. One effort was made in the late 1970s but lasted only 3 years. The editors terminated the "publication" in 1980 with a paraphrase of Lincoln Steffens' pronouncement on the Russian revolution: "I have been over into the future and it works" (238:250). They said: "We have visited the future and it doesn't work" (67). Cost was recognized as a factor, but the primary problem was the unwillingness of authors to publish in a format which they regarded as having limited readership and lack of status.

Critical mass, the commitment of enough publishers to the medium, is, of course, a problem, as Bernal recognized when he predicted in his proposal to distribute separates that: "If it is not practically all-inclusive it will fail" (32:449). Other problems have also been raised. Electronic journals may change the technology of storing and transmitting scientific papers but may not radically change any of the intellectual, social, ethical, or other problems. The primary reasons for lag time may not reside in the technology; the process of authoring, editorial processing, and reviewing may still take as long as before. Each communication must still take its place in a queue with others for consideration and processing. Questions about readability in electronic formats have also been raised. Some studies have shown that fatigue generally occurs after only 3 minutes of continuous reading of texts on computer monitors (360). One author commented: "2000 words may read beautifully in print, but put them on the screen and they are a colossal bore" (101).

Acceptability of electronic forms of publishing by authors and readers remains a primary issue, but readership, status, and readability are not the only concerns. The very flexibility or malleability of computer-stored information which is one of its chief assets may also be one of its limiting factors. What happens to the concept of "edition," a validated and fixed statement made by a particular individual at a particular point in time, which can be identified and referred to, and which is a fundamental element of scholarship? Swanson called attention forcibly to this issue in his review of a conference on *The Role of the Library in an Electronic Society:*

> I find the idea of an electronic journal troubling. Criticism is essential to the growth of knowledge; the purpose of publishing is to give ideas a concrete form available to public scrutiny. I do not think that I would be willing to extend, build on, critique, or even cite any article available only in computer storage unless I felt confident that the article would not be changed after my own work was made public. (427:318)

Electronic storage eliminates the need to replicate information, but, historically, recorded information has survived largely by virtue of having been available in multiple copies (325).

Concerns have also been expressed about the threat to the rewards and incentives to authors inherent in our current scientific communication systems. Even Lederberg, after describing the contributions that the electronic media can make to scholarship, says: ". . . no piece of work, no claim to priority, is authentically recorded until it has appeared in public print in a respectable refereed journal" (273). The new technology will, nevertheless, inevitably lead to adoption of unconventional formats. Responses to a questionnaire sent to 226 American institutions of higher education with graduate programs in sociology revealed that 55 percent thought they would accept electronic publication as equivalent to print if all other factors were held constant, but not a single respondent thought it was superior (398). Computers are providing a powerful adjunct to print technology, but at this point it is difficult to predict what their role will eventually be. The following two statements by Starr seem to describe an evolving consensus: "The printed page has such obvious advantages in ease of access and flexibility of use that there is no likelihood of its abandonment for most ordinary reading. . . . A primary use of the new technology will be to deliver printed texts more efficiently than before" (420:144).

Future Developments

It is more likely that the impact of the new technology, rather than in the form of the electronic journal, will be demonstrated in new modes and techniques of synthesizing, organizing, presenting, and disseminating information. Among the pioneering examples are the National Library of Medicine's "Hepatitis Knowledge Base" and the *ISI Atlas of Science* to which I referred in Chapter 4. They may point to one way in which a current consensus in a subject can be made readily available with the aid of the computer. The technology actually played a relatively small role in a process in which the designers recognized that the "tasks are more intellectual than mechanical" (38:171).

Computers were used in several ways in compiling the "Hepatitis Knowledge Base": (i) for information retrieval to identify the core literature, (ii) for text entry, (iii) for consensus formation by using an experimental "computer-conferencing system" in which the experts were able to communicate with one another, and (iv) to provide online access to the database to users. The investigators speculated on whether such a technique could be expanded to constitute a "living textbook," but considering what it must have cost to provide even this small segment of a total knowledge base, it remains highly conjectural.

Some of the new roles of the computer in society are described by Hiltz and Turoff in a book published in 1978 called the *Network Nation*. It envisions individuals communicating and receiving information largely through "computer conferencing systems." This is a generic term for systems which, in their words, use:

... the computer to structure, store and process written communications among a group of persons. When something is entered through one terminal it may be obtained on the recipients' terminals immediately or at any time in the future until it is purged from the computer's memory. (217:7)

The authors regard these systems as having great potential for increasing and facilitating communication among all kinds of people, including scientists and scholars. In using the system participants type their messages and contributions on a terminal which transmits them to a central computer to which other authorized terminals have access. Participants can enter or receive messages at any time, and they can be made available to as many or as few other participants as the sender desires. The cost of participating in such a system was estimated by the authors even in 1978 as being under $8 an hour, regarded then as competitive with long-distance telephone rates.

They see the system being used by scientists in a number of ways:

1. *Messages* which can be addressed to any or all members of any group of participants.
2. *Conferences* in which designated participants can enter or leave a discussion on a topic at any time. They may last anywhere from a week to several months, and transcripts can be produced for the record if desired.
3. *"Notebooks"* which are private online work spaces that can also be made available to others. This space can be used for drafts or preprints of papers, or for co-authoring articles or joint research proposals.
4. *"Bulletins"* which are public space and require the management participation of organizers or editors. This component most closely resembles our current print media, or the proposed electronic journals, and would have to develop similar marketing and control procedures.

The authors see several kinds of electronic journals developing out of such systems.

Many advantages are seen in using computer conferencing systems. They can document all exchanges between scientists when such documentation is desired. They can facilitate bringing people together. Lederberg comments on the difficulties of arranging for conference calls among busy individuals: ". . . several weeks prior notice (or other rigid prearrangements) has been needed to schedule teleconferences if four or more people were required simultaneously" (273:1315). Lederberg also points out that in computer conferencing systems the respondent "is liberated from the tyranny of synchronizing his own mental processes to those of an external actor" (273:1315).

Experiments with such systems have been in progress for several years. Lederberg described the Stanford University Medical Experimental Computer-Artificial Intelligence Medical (SUMER-AIM) project in 1978 (273). Hiltz and Turoff's book is based to a large extent on their experiences with the Electronic Information Exchange System (EIES) which has been available to a number of participants and which was used by the National Library of Medicine for the development of its "Hepatitis Knowledge Base."

Hiltz and Turoff recognize that these systems have problems and limitations. Conveying information by voice contact and visual observation adds a dimension which is not present in digitalized transmissions. The authors found that in some cases computerized conferencing did not eliminate the desire for face to face encounter but in fact increased it. Rather than cutting down on travel, which is one of the goals of these systems, in some cases participating members identified others at a distance who had common problems and whom they wished to visit.

The authors of the *Network Nation* made the prediction in 1978 that: "Computerized conferencing will be a prominent form of communication in most organizations by the mid-1980's." We are still some distance from these goals. At this point we are in a highly transitional period in the development of our information processing and disseminating systems in science. The situation cannot be said to have changed considerably since 1977, when Manten said:

> We are likely to enter a period of instability in scientific and technological information transfer. It will take some time before things settle down again in some configuration different from the present. (292)

Since the beginnings of scientific journalism, the journal has served a dual role, as a vehicle to disseminate information and as an archive in which it is stored against future use. It is becoming more and more apparent that these are incompatible roles and that ". . . the traditional journal is not a highly efficient vehicle for disseminating information to the individual scientist, who reads very few of the papers in any given issue, according to repeated studies" (53). As scientists become more specialized, the markets for journals shrink, which contributes to their costliness. Creation of new journals also tends to increase the volume of literature. Henry Fielding in his *History of Tom Jones* was referring to literary journals, two of which he edited in the eighteenth century, but the analogy can be with some justice applied to scientific journals as well.

> Such histories as these do, in reality, much resemble a newspaper which consists of just the same number of words, whether there be any news or not. They may likewise be compared to a stagecoach which performs constantly the same course, empty as well as full. (137)

The issues are not so much between electronic and conventional forms of journals but between selective and generalized distribution of information. The mechanisms for the management and distribution of separates may have been cumbersome and unwieldy when Bernal made his proposal, but the new technology provides us with an opportunity to rethink the process.

The contributions which technology can make are nevertheless limited. The computer can process and store words at incredible speeds, but choosing the right words to articulate an idea and being able to interpret the words correctly are still individual intellectual efforts which technology does little to change. Memory and the associated retrieval techniques still depend on cognitive association trails which must be articulated for machine as well as manual systems. Not all information can be transmitted through print or electronic media. Collins comments on the limitations of any kind of discourse, oral, written, or machine mediated:

> ... it was found that a scientist who was to learn to build a successful copy of a laser nearly always needed to spend sometime in close interaction with another who had already built one. (82:207)

Machines, says another author, cannot as yet substitute for the cognitive processes involved in the analysis of a question, design of an overall strategy, choice of an appropriate information source, or selection of acceptable search terms (25:144). It is also not clear what machines can do to improve the quality of scientific information. In fact, it has been suggested that the major improvement which could be made in scientific journals is to place more importance on developing communications skills and raising research standards.

Nevertheless, it is also apparent that the new technologies are providing us with increasing access to large amounts of information and the ability to identify relevant portions quickly. We can look forward to an increase in the number of handbooks and data tables online to replace the manuals and reference data collections that need constantly to be updated. Expanded database storage and improvements in network communication should give the most remote user the same kind of access to vast information stores that is available to those in the most populous areas.

Questions have been raised about whether print media will be used in the future except in the same way scholars today study clay tablets and papyrus rolls. Most observers are confident, however, that print on paper will survive just as other communication technologies have survived in the face of change. In fact, the introduction of newer media frequently has the effect of increasing the use of the older media. In 1950 when television began to appear, there were 11,022 books published in the United States. In 1970 when computers began to make their entry into information dissemination, and after 20 years of television, the number had risen to 36,071. In 1979 it was 45,182 (267:85). The retort made to an apostle of the "paperless society" that "... the more computer programmers there are, the more books will be needed on computer programming" proved to be prophetic (259:132). In 1982, *Publishers Weekly* printed a list of almost 250 current titles on computers (105). Two years later, a directory of 1,500 pages appeared which "provides access to over 20,000 current U.S. and foreign books, serials and pamphlets—many of which were published within the last year" (83).

Despite predictions of the imminent demise of the scientific journal, it has continued to proliferate. *Nature* publishes a highly selective review of new English-language scientific journals every year. There were 62 in 1981 (328) and

124 in 1982 (327). Sociologists have made clear that the scientific journal over the past 300 years has evolved as a social institution and a focal element in the control system of science, as well as a way of disseminating information. Social institutions like the family and marriage may be modified by the introduction of new technology, but the "definite, continuous and organized patterns of behavior" and the "normative ordering and regulation" (125) tend to persist.

The emphasis in this book has been on the standard and conventional methods of organizing, storing, and retrieving information. Many of these aspects will not change no matter what new technology is introduced. It will still be important for the information system user to be aware of the distinction between controlled and natural languages. At the same time we should be aware of and knowledgeable in using the new technology. The kind of technology to be used will eventually probably be decided on a cost-effectiveness basis. In spite of its recognized problems and deficiencies, the current scientific information systems have made a great contribution to scientific progress. It is essential for every scientist and practitioner to learn how to use them well.

Appendix

A Selected List of Guides
to the Literature

Blake JB, Roos C, eds. *Medical reference works, 1679–1966.* Chicago: Medical Library Association, 1967. Suppl. 1–3, 1970–1974.

Blanchard JR, Farrell L, eds. *Guide to sources for agricultural and biological research.* Berkeley: University of California Press, 1981.

Bottle RT, Wyatt HW, eds. *The use of biological literature,* 2nd ed. Hamden, Conn.: Archon Books, 1972.

Bottle RT, ed. *The use of chemical literature,* 3rd ed. London: Butterworths, 1979.

Brunn AL. *How to find out in pharmacy; a guide to sources of pharmaceutical information.* New York: Pergamon, 1969.

Chen C. *Health sciences information sources.* Cambridge, Mass.: MIT Press, 1981.

Chen C. *Scientific and technical information sources.* Cambridge, Mass.: MIT Press, 1977.

Davis EB. *Using the biological literature; a practical guide.* New York: Dekker, 1981.

Grogan D. *Science and technology; an introduction to the literature,* 4th ed. London: Cline Bingley, 1982.

Jenkins FB. *Science reference sources,* 5th ed. Cambridge, Mass.: MIT Press, 1969.

Lasworth EJ. *Reference sources in science and technology.* Metuchen, N.J.: Scarecrow Press, 1972.

Londos EG. *Compendium of current source materials for drugs.* Metuchen, N.J.: Scarecrow Press, 1982.

Maizell RE. *How to find chemical information; a guide for practicing chemists, teachers, and students.* New York: Wiley, 1979.

Malinowsky HR, Richardson JM. *Science and engineering literature; a guide to reference sources,* 3rd ed. Littleton, Colo.: Libraries Unlimited, 1980.

Mellon MC. *Chemical publications,* 5th ed. New York: McGraw-Hill, 1982.

Morton LT, Godbolt S, eds. *Information sources in the medical sciences,* 3rd ed. London: Butterworths, 1984.

Roper FW, Boorkman JA. *Introduction to reference sources in the health sciences.* Chicago: Medical Library Association, 1980.

Sewell W. *Guide to drug information.* Hamilton, Ill.: Drug Intelligence Publications, 1976.

Sheehy EP, ed. *Guide to reference books,* 9th ed. Chicago: American Library Association, 1976. Suppl. 1–2, 1980–82.

189

Smith RC, Reid WM. *Guide to the literature of the life sciences,* 9th ed. Minneapolis: Burgess Publishing Co., 1980.

Subramanyan K. *Scientific and technical information resources.* New York: Dekker, 1981.

Thornton JL. *Medical books, libraries and collectors,* 2nd ed. London: Deutsch, 1966.

Thornton JL. *Scientific books, libraries and collectors,* 3rd ed. London: Library Association, 1971.

Walford AJ, ed. *Guide to reference material. Vol. 1. Science and technology,* 4th ed. London: Library Association, 1980.

Wexler P. *Information sources in toxicology.* New York: Elsevier/North Holland, 1982.

Woodburn HM. *Using the chemical literature; a practical guide.* New York: Dekker, 1974.

References

1. AAP to develop electronic publishing standards. *Bull Am Soc Inf Sci* 1983, 8(8)6.
2. Abelson PH. The editing of science. *Science* 1971, 171:1101.
3. Abelson PH. Electronics and scientific communication. *Science* 1980, 210 (Oct 17) Editorial.
4. Abelson PH. Information exchange groups. *Science* 1966, 154 (Nov 11) Editorial.
5. Adair RK. A physics editor comments on Peters and Ceci's peer review study. *Behav Brain Sci* 1982, 5:196.
6. Adair WC. Citation indexes for scientific literature? *Am Doc* 1955, 6:31–2.
7. Adams S. The review literature of medicine. Preface to: *Bibliography of Medical Reviews, volume 6: Cumulation 1955–1961*. Washington, D.C.: National Library of Medicine, 1961.
8. Adams S, Baker DB. Mission and discipline orientation in scientific abstracting and indexing services. *Libr Trends* 1968, 16:307–22.
9. Adler MJ. *How to read a book: the art of getting a liberal education.* New York: Simon and Schuster, 1940.
10. Allen WE. Respositories for scientific publication. *Science* 1922, 56:197–8.
11. *Almanach für Aertze und Nichtaerzte für die Jahre. 1783.* Jena, 1783.
12. Altman DG. Statistics and ethics in medical research. *Br Med J* 1980, 281:1182–4, 1267–9, 1336–8, 1399–401, 1473–5, 1542–4, 1612–4; 1981, 282:44–7.
13. Altman L, Melcher L. Fraud in science. *Br Med J.* 1983, 286:2003–6.
14. American Association of Medical Colleges. *President's Activity* Report 82-15. 15 April 1982.
15. American Chemical Society. *CAS today: facts and figures about Chemical Abstracts service.* Columbus, 1980.
16. American National Standards Institute, Inc. *American National Standard for Writing Abstracts,* ANSI Z39.14-1979. New York, 1979.
17. Anderson VL, McLean RA. *Design of experiments: a realistic approach.* New York: Dekker, 1974.
18. Asher R. Making sense. *Lancet* 1959, 2:359–65.
19. Bachrach A. Concerning non-serial proceedings. *NLM Tech Bull,* No. 161, Sept 1982:7–8.
20. Baker DB, Horiszny JW, Metanomski WV. History of abstracting at Chemical Abstracts Service. *J Chem Inf Comput Sci* 1980, 20:193–201.
21. Barden W. The appliance computer. *Pop Computing,* 1982, 2(9):58–62.
22. Barlup J. Relevancy of cited articles in citation indexing. *Bull Med Libr Assoc* 1969, 57:260–3.
23. Barr KP. Estimates of the number of currently available scientific and technical periodicals. *J Doc* 1967, 23:110–6.

191

24. Bartlett LC. Filed and found: a personal information storage and retrieval system. *J Am Med Assoc* 1967, 199:244–5, 250, 264.
25. Bates MJ. Search techniques. *Annu Rev Inf Sci* 1981, 16:139–69.
26. Bean WB. Tower of Babel. *Arch Intern Med* 1962, 110:375–81.
27. Belkin NJ, Oddy RN, Brooks HM. Ask for information retrieval: Part 1. Background and theory. *J Doc* 1982, 38:61–71.
28. Bell D. The measurement of knowledge. *In:* Sheldon EB, Moore WE, eds. *Indicators of social change.* New York: Russell Sage Foundation, 1968, pp. 145–246.
29. Ben-David J. Organization, social control, and cognitive change in science. *In:* Ben-David J, Clark T, eds. *Culture and its creators.* Chicago: University of Chicago Press, 1972, p. 244–323.
30. Ben-David J. The universities and the growth of science in Germany and the United States. *Minerva* 1968/69, 7:1–35.
31. Bergeijk D, Risseeuw M. The language barrier in the dissemination of scientific information and the role of the ITC. *J Inf Sci* 1980, 2:37–42.
32. Bernal JD. Project for scientific publication and bibliography. In: *The social function of science.* London: Routledge, 1939, pp. 449–55.
33. Bernal JD. Provisional scheme for central distribution of scientific publication. In: Royal Society Scientific Information Conference, 21 June–2 July 1948. *Reports and papers submitted.* London: Royal Society, 1948, pp. 253–7.
34. Bernal JD. The transmission of scientific information: a user's analysis. In: *Proceedings of the International Conference on Scientific Information, Washington, D.C. 1958.* Washington, D.C.: National Academy of Sciences, 1959, pp. 76–95.
35. Bernard HR. Computer-assisted referee selection as a means of reducing potential editorial bias. *Behav Brain Sci* 1982, 5:202.
36. Bernfeld GE, Shalter MD. The question of authorship. *Med Commun* 1982, 10:7–10.
37. Bernstein J. Topless in Hamburg. *Am Scholar* 1980/81, 50:7–14.
38. Bernstein LM et al. The hepatitis knowledge base (short form). *Ann Intern Med* 1980, 93:166–81.
39. Berry EM. The evolution of scientific and medical journals. *N Engl J Med* 1981, 305:400–2.
40. Beyer JM, Sikorski SM. *Preliminary report: results of a survey of editors of ten leading journals in each of four scientific fields.* Buffalo: State University of New York, School of Management, 1975 (Working paper no. 226).
41. Billings JS. *Selected papers, compiled with a life of Billings by FB Rogers.* Chicago: Medical Library Association, 1965.
42. BioSciences Information Service. *Serial sources for the Biosis database.* Philadelphia, 1980.
43. Birch BJ. Tracking down translations. *ASLIB Proc* 1979, 31:500–11.
44. Blackwelder RE. *Taxonomy, a text and reference book.* New York: John Wiley, 1967.
45. Bogdanove EM. Regulation of TSH secretion. *Fed Proc* 1962, 21:623–7.
46. Bohr N. *Atomic physics and human knowledge.* New York: Science Editions, 1961, pp. 67–8.
47. Bollum FJ. Trooping the harem (a review of *Molecular and Cellular Biology* and other related journals). *Nature* 1982, 299:497.
48. Booth CC. Medical communication: the old and the new. *Br Med J* 1982, 285:105–8.

49. Borko H, Bernier CL. *Abstracting concepts and methods.* New York: Academic Press, 1975.

50. Bottle RT. Information obtainable from analyses of scientific bibliographies. *Libr Trends* 1973, 22:60–71.

51. Bottle RT, Wyatt HV, eds. *The use of biological literature,* 2nd ed. Hamden, Conn.: Archeon Books, 1971.

52. Bowen CD. *Francis Bacon, the temper of a man.* Boston: Little, Brown, 1963.

53. Bowen DHM. Member subscriptions. In: Council of Biology Editors. *Economics of scientific journals.* Bethesda, Md.: Council of Biology Editors, 1982, pp. 1–4.

54. Bradford SC. *Documentation.* Washington, D.C.: Washington Affairs Press, 1950.

55. Branscomb LM. The electronic library. *J Commun* 1981, 31:143–50.

56. Branscomb LM. Information: the ultimate frontier. *Science* 1979, 203:143–7.

57. Broad W, Wade N. *Betrayers of the truth.* New York: Simon and Schuster, 1982.

58. Broad WJ. Imbroglio at Yale. *Science* 1980, 210:38–41, 171–3.

59. Broad WJ. The publishing game: getting more for less. *Science* 1981, 211:1137–9.

60. Broadhurst PL. Coordinate indexing: a bibliographic aid. *Am Psychol* 1962, 17:137–42.

61. Brodman E. *The development of medical bibliography.* Baltimore: Medical Library Association, 1954.

62. Brodman E, Taine SI. Current medical literature: a quantitative survey of articles and journals. In: *Proceedings of the International Conference on Scientific Information.* Washington, D.C.: National Academy of Sciences, 1959, 1:121–33.

63. Bronowski J. Principles of tolerance. *Atl Mon* 1973, 232(6):60.

64. Burton RE, Kebler RW. The "half-life" of some scientific and technical literature. *Am Doc* 1960, 11:18–22.

65. Bush V. As we may think. *Atl Mon* 1945, 176(1):101–8.

66. Calvin WH. A computer-assisted literature reference system. *Comput Programs Biomed* 1972, 2:291–6.

67. Campbell R. Investing in the future? *New Sci* 1982, 94(8 Apr):99.

68. Campbell, R. Making sense of journal publishing. *Nature* 1982, 299:491–2.

69. Carpenter MP, Narin F. The adequacy of the *Science Citation Index* (SCI) as an indicator of international activity. *J Am Soc Inf Sci* 1981, 32:430–9.

70. Carpenter MP, Narin F. The subject composition of the world's scientific journals. *Scientometrics* 1980, 2:53–63.

71. Carrol KH. An analytical survey of virology literature in two announcement journals. *Am Doc* 1969, 20:234–7.

72. Carson J, Wyatt HV. Delays in the literature of medical microbiology: before and after publication. *J Doc* 1983, 39:155–65.

73. Charen T. *Medlars indexing manual.* Bethesda, Md.: National Library of Medicine, 1981, p. 4.1.

74. Chargaff E. Building the Tower of Babel. *Nature* 1974, 248:776–9.

75. Chase JM. Normative criteria for scientific publication. *Am Sociol* 1970, 5:262–5.

76. Chernin E. First do no harm. In: Warren KS, ed. *Coping with the biomedical literature.* New York: Praeger, 1981, pp. 49–65.

77. Clarke BL. Multiple authorship trends in scientific papers. *Science* 1964, 143:822–4.

78. Coblans H. New methods and techniques for the communication of knowledge. *UNESCO Bull Libr* 1958, 11(7):154–75.

79. Cole FJ, Eales NB. The history of comparative anatomy. *Sci Prog* 1917, 11:578–96.

80. Cole JR, Cole S. *Social stratification in science.* Chicago: University of Chicago Press, 1973.
81. Cole S, Cole JR. Scientific output and recognition: a study in the operation of the reward system in science. *Am Sociol Rev* 1967, 32:377–90.
82. Collins HM. The seven sexes: a study in the sociology of phenomenon; or the replication of experiments in physics. *Sociology* 1975, 9:205–24.
83. *Computer books and serials in print.* New York: Bowker, 1984.
84. Conant JB. *Science and common sense.* New Haven: Yale University Press, 1951.
85. Confrey EA. Information exchange groups to be discontinued. *Science* 1966, 154:843.
86. Congrat-Butlar S, ed. *Translations and translators.* New York: Bowker, 1979.
87. Consequences mortelle d'une erreur typographique. *Med dans le Monde* (Suppl. *Semaine Med*) 1950, 26:537.
88. Cooper WS. Is interindexer consistency a hobgoblin? *Am Doc* 1969, 20:268–78.
89. Corning ME. International biomedical communications: the role of the United States National Library of Medicine. *Health Commun* 1980, 6:212–42.
90. Costello WY. *The scholarly curriculum at early seventeenth century Cambridge.* Cambridge, Mass.: Harvard University Press, 1958.
91. Council of Biology Editors Style Manual Committee. *Style manual: a guide for authors, editors, and publishers in the biological sciences,* 5th ed. Bethesda, Md.: Council of Biology Editors, 1983.
92. Cournand A. Science in service of society. *The Sciences* 1981, 21(8):7–9.
93. Crane D. The gatekeepers of science: some factors affecting the selection of articles for scientific journals. *Am Psychol* 1967, 2:195–201.
94. Creager RJ. Medical literature filing system in family practice residency program. *J Fam Prac* 1983, 16:621–4.
95. Cremmins E. *The art of abstracting.* Philadelphia: ISI Press, 1982.
96. Crick F. *The double helix:* a personal view. *Nature* 1974, 248:766–9.
97. Cronin B. The need for a theory of citing. *J Doc* 1981, 37:16–24.
98. Crowson RA. *Classification and biology.* Chicago: Aldine Publishing Co., 1971.
99. Cuadra RN et al., eds. *Directory of Online Databases.* Santa Monica, Calif.: Cuadra Associates, Inc., 1982, vol. 4, p. 100.
100. Culliton BJ. The academic-industrial complex. *Science* 1982, 216:960–2.
101. Dahlen R. Electronic publications: steps forward—and back. *Publ Weekly* 1982, 22(22):26.
102. Dan B. The paper chase. *J Am Med Assoc* 1983, 249:2872–3.
103. Darwin C. *The life and letters of Charles Darwin.* New York: Appleton, 1896.
104. Davis J. What's new in biotechnology? *Nature* 1982, 299:493–4.
105. Davis J, Goldstein W. Books on computers: a current checklist. *Pub Weekly* 1982, 222(20):29–49.
106. Davis RM. Computers and electronics for individual services. *Science* 1982, 215:852–5.
107. Davis W. Microfilms make information accessible. In: Bernal JD. *The social function of science.* London: Routledge, 1939, pp. 455–7.
108. Day RA. How to write a scientific paper. *ASM News* 1975, 41:486–94.
109. Day RA. *How to write and publish a scientific paper,* 2nd ed. Philadelphia: ISI Press, 1983.
110. De Alarcon R. A personal medical reference index. *Lancet* 1969, 1:301–5.
111. DeBakey L. *The scientific journal editorial policies and practices: guidelines for editors, reviewers and authors.* St. Louis: Mosby, 1976.

112. DeBakey L, DeBakey S. Ethics and etiquette in biomedical communication. *Perspect Biol Med* 1975, 18:522–40.

113. De Grazia A. The scientific reception system and Dr. Velikovsky. *Am Behav Sci* 1963, 71:38–56.

114. *Directory of online databases.* Spring, 4(3). Santa Monica, Calif.: Cuadra Associates, 1983.

115. *Directory of publishing opportunities in journals and periodicals,* 5th ed. Chicago: Marquis Academic Media, 1981.

116. Dobell C. Dr. O. Uplavici (1887–1938). *Parasitology* 1938, 30:239–41.

117. Dolby RGA. The transmission of science. *Hist Sci* 1977, 15:1–43.

118. Drake S. *The unsung journalist and the origins of the telescope.* Los Angeles: Zeitlin and Ver Brugge, 1976.

119. Dray S. Information exchange group no. 5. *Science* 1966, 153:694–5.

120. Dudley H. *The presentation of original work in medicine and biology.* Edinburgh: Churchill-Livingstone, 1977, p. 19.

121. Durack DT. The weight of medical knowledge. *N Engl J Med* 1978, 298:773–5.

122. Dym ED, Shirey DL. A statistical decision model for periodical selection for a specialized information center. *J Am Soc Inf Sci* 1973, 24:110–9.

123. *Ecclesiastes,* 12:12–4.

124. Egeland J. The SUNY Biomedical Communications Network: six years of progress in on-line bibliographic retrieval. *Bull Med Libr Assoc* 1975, 63:189–94.

125. Eisenstadt SN. Social institutions. In: Sills DL, ed. *International encyclopedia of the social sciences.* New York: Macmillan, 1968, 14:409–29.

126. Elkana Y et al., eds. *Toward a metric of science: the advent of science indicators.* New York: Wiley, 1977.

127. Elsevier Company. Publisher's announcement, 1983.

128. Emard JP. An interview with BRS' William Marovitz. *Online* 1983, 7(3):15–9.

129. *Encyclopedia of information systems and services,* 4th ed. Detroit: Gale Research Co., 1981.

130. Espinasse M. *Robert Hooke.* London: Heinemann, 1956.

131. *Essais et observations de médécine de la Société d'Edimbourg.* (Paris) 1740, 1:vi.

132. Fabregé AC. Open information and secrecy in research. *Perspect Biol Med* 1982, 25:263–78.

133. Fawcett PJ. Personal filing systems revisited. *Ear Nose Throat J* 1978, 59:82–9.

134. Feinstein AR. *Clinical biostatistics.* St. Louis: Mosby, 1977.

135. Feinstein AR. Clinical biostatistics. XXV. A survey of the statistical procedures in general medical journals. *Clin Pharmacol Ther* 1974, 15:97–107.

136. Felig P. Retraction. *Diabetes* 1980, 29:672.

137. Fielding H. *The history of Tom Jones.* New York: Modern Library, 1940, p. 40.

138. Fink JL, Gerbino PP. Co-authorship of scientific papers. *Drug Intell Clin Pharm* 1977, 11:558.

139. Fluge PF. A modern Icarus reaches to the sun for power to fly. *Smithsonian* 1981, 11(Feb):72.

140. Foster GM, Anderson BG. *Medical anthropology.* New York: Wiley, 1978, p. 23.

141. Foster M. *Lectures on the history of physiology during the sixteenth, seventeenth and eighteenth centuries.* Cambridge: Cambridge University Press, 1901.

142. Frank RG Jr. Science, medicine and the universities of early modern England. *Hist Sci* 1973, 11:194–216, 239–69.

143. Fry BM, White HS. *Economics and interaction of the publisher-library relationship in the production and use of scholarly and research journals.* Washington, D.C.: National Science Foundation, 1976.

144. Fuller EA. A system for filing medical literature. *Ann Intern Med* 1968, 68:684–93.

145. Future computer storage technologies. *Libr Systems Newsl* 1983, 3(12):92–5.

146. Gaeke RF, Gaeke MEB. Filing medical literature: a textbook-integrated system. *Ann Intern Med* 1973, 78:985–7.

147. Gardner JL. The conference as an integral component in the science and technology dissemination network. In: Zamara G, Adamson MC, eds. *Conference literature: its role in the dissemination of information.* Marlton N.J.: Learned Information Inc., 1981, p. 8.

148. Garfield E. Can citation indexing be automated. In: Stevens ME, ed. *Methods for mechanized documentation.* Washington, D.C.: National Bureau of Standards, 1965, p. 189. Cited by Smith (414).

149. Garfield E. Citation indexes in sociological and historical research. *Am Doc* 1963, 14:289–91.

150. Garfield E. *Citation indexing, its theory and application in science, technology and humanities.* New York: Wiley, 1979.

151. Garfield E. Citations. *New Sci* 1968, 39:565–6.

152. Garfield E. *Essays of an information scientist* (referred to subsequently as *Essays*), vol. 1–5. Philadelphia: ISI Press, 1977–1983.

153. Garfield E. ABC's of cluster mapping. In: *Essays,* 4:634–49.

154. Garfield E. Alternative forms of scientific publication. In: *Essays,* 4:264–8.

155. Garfield E. Anonymity in refereeing? Maybe. But anonymity in authorship? No! In: *Essays,* 2:438–40.

156. Garfield E. Bradford's law and related statistical patterns. In: *Essays,* 4:476–83.

157. Garfield E. Can machines be scientific translators? In: *Essays,* 4:574–8.

158. Garfield E. Do French scientists who publish outside France and/or in English do better research. In: *Essays,* 3:498–503.

159. Garfield E. The ethics of scientific publication. In: *Essays,* 3:644–51.

160. Garfield E. From citation awareness to bibliographic amnesia. In: *Essays,* 4:503–7.

161. Garfield E. Highly cited articles. 39. Biochemistry papers published in the 1950's. In: *Essays,* 3:147–54.

162. Garfield E. More on the ethics of scientific publication. In: *Essays,* 5:621–6.

163. Garfield E. The mystery of the transposed journal lists—wherein Bradford's Law of Scattering is generalized according to Garfield's Law of Concentration. In: *Essays,* 1:222–23.

164. Garfield E. The 'obliteration phenomenon' in science . . . and the advantage of being obliterated. In: *Essays,* 2:396–8.

165. Garfield E. Of conferences and reviews. In: *Essays,* 2:286–389.

166. Garfield E. Proposal for a new profession; scientific review. In: *Essays,* 3:84–7.

167. Garfield E. Scientometrics comes of age. In: *Essays,* 4:313–8.

168. Garfield E. The 300 most-cited authors, 1961–1976, including co-authors. 3A. Their most cited papers. Introduction and journal analysis. In: *Essays* 3:689–700.

169. Garfield E. Uncitedness III: the importance of not being cited. In: *Essays,* 1:413–4.

170. Garfield E. Is information retrieval in the arts and humanities inherently different from that in science? *Libr Q* 1980, 50:40–57.

171. Garfield E. Preface. In: *Guide and Journal Lists: Index to Scientific Reviews 1974.* Philadelphia: Institute for Scientific Information, 1974.
172. Garfield E. *Science Citation Index Annual:* a bibliometric analysis of references processed for the 1974 Science Citation Index. In: *Journal Citation Reports.* Philadelphia: Institute for Scientific Information, 1976.
173. Garfield E. Significant journals of science. *Nature* 1976, 264:609–15.
174. Garfield E. Trends in biochemical literature. *Trends Biochem Sci* 1979, 4(12): N290–5.
175. Garvey MD, Griffith BC. Scientific communication: its role in the conduct of research and creation of knowledge. *Am Psychol* 1971, 26:349–62.
176. Gasking EB. Why was Mendel's work ignored. *J Hist Ideas* 1959, 20:60–84.
177. Gaston J. *The reward system in British and American science.* New York, Wiley, 1978.
178. Gehlbach SH. *Interpreting the medical literature: a clinician's guide.* Lexington, Mass.: Collamore Press, 1982.
179. Geiseler PJ. Unnecessary duplication of publication. *N Engl J Med* 1980, 302:1209–10.
180. Giamatti AB. The university, industry and cooperative research. *Science* 1982, 218:1278–80.
181. Gilbert GN. The transformation of research findings into scientific knowledge. *Soc Stud Sci* 1976, 6:281–306.
182. Gillespie CC. *Society and polity in France at the end of the Old Regime.* Princeton N.J.: Princeton University Press, 1980.
183. Gillon R. The function of criticism. *Br Med J* 1981, 283:1633–9.
184. Glantz SA. Biostatistics: how to detect, correct and prevent error in the medical literature. *Circulation* 1980, 61:1–7.
185. Goffman W. The ecology of the biomedical literature and information retrieval. In: Warren KS, ed. *Coping with the biomedical literature.* New York: Praeger, 1981, pp. 31–46.
186. Goldblatt H. Die neuere Richtung der experimentellen Rachitisforschung. *Ergeb Allg Pathol Pathol Anat* 1931, 25:57–249.
187. Goldsmith M. *Sage: a life of JD Bernal.* London: Hutchinson, 1980.
188. Goldstein CM. Optical disk technology and information. *Science* 1982, 215:862–8.
189. Gore SM, Jones IG, Rytter EC. Misuse of statistical methods: critical assessment of articles in BMJ from January to March 1976. *Br Med J* 1977, 1:85–7.
190. Gottschalk CM, Desmond WF. Worldwide census of scientific and technical serials. *Am Doc* 1963, 14:188–94.
191. Gould SJ. The Piltdown conspiracy. In: *Hen's teeth and horse's toes.* New York: Norton, 1983, p. 201–26.
192. Greene JC. The Kuhnian paradigm and the Darwinian revolution in natural history. In: Roller D, ed. *Perspectives in the history of science and technology.* Norman: University of Oklahoma Press, 1971, p. 3–25.
193. Greenstein JS. Studies on a new, peerless contraceptive agent; a preliminary, final report. *Can Med Assoc J* 1965, 93:1351–5.
194. Griffith BC, Servi PN, Ander AL, Drott MC. The aging of scientific literature: a citation analysis. *J Doc* 1979, 35:179–96.
195. Griffith BC, ed. Introduction. *Key papers in information science.* White Plains, N.Y.: Knowledge Industries, 1980.

196. Griner LA, Hutton NE. Pathology data retrieval system adopted at a zoological garden. *J Am Vet Med Assoc* 1968, 153:885–92.
197. Guerlac H. A lost memoir of Lavoisier. *Isis* 1959, 50:125–9.
198. *Guide to the Indexes.* Philadelphia, Biosciences Information Service, 1972.
199. Gutheil TG. Rapid retrieval of literature for the practicing psychiatrist: a practical approach. *Am J Psychiat* 1974, 131:1145–6.
200. Haas WJ. Computing in documentation and scholarly research. *Science* 1982, 215:857–61.
201. Hagstrom WO. *The scientific community.* New York: Basic Books, 1965.
201a. Hall A, Hall MB, eds. *The correspondence of Henry Oldenburg.* Madison: University of Wisconsin Press, 1965–77. 11 v.
202. Hall MB. Science in the early Royal Society. In: Crossland M, ed. *The emergence of science in western Europe.* New York: Science History Press, 1975.
203. Hamblin TJ. Fake. *Br Med J* 1981, 283:1671–4.
204. *Handbook of physiology.* Washington, D.C.: American Physiological Society, 1959. Foreword to Section 1, vol. 1, p. v.
205. Hannaway CF. *Medicine and public welfare and the state in eighteenth century France: the Société Royale de Médecine of Paris (1776–1793).* Ph.D. Dissertation, Johns Hopkins University, 1974.
206. Harper F. *The Code of Hammurabi, King of Babylon,* 2nd ed. Chicago: University of Chicago Press, 1904.
207. Hartner EP. *An introduction to automated literature searching.* New York: Dekker, 1981.
208. Hartree EF. Ethics for authors: a case history of acrosin. *Perspect Biol Med* 1976, 20:82–91.
209. Harvey AM, McKusick VA, eds. *Osler's textbook revisited.* New York: Appleton-Century-Crofts, 1967.
210. Hawke DH. *Franklin.* New York: Harper, 1976.
211. Hawkins DT. Online bibliographic search strategy developments. *Online* 1982, 6:12–9.
212. Hellmers H. A simple and efficient file system for reprints. *Biosciences* 1964, 14:24.
213. Helmholtz H. On thought in medicine. In: *Popular lectures on scientific subjects.* London: Longmans, Green & Co., 1893, vol. 2, pp. 199–236.
214. Helmholtz H. *Popular lectures on scientific subjects.* New York: Appleton, 1895, p. 12–3.
215. Hickey T. The journal in the year 2000. *Wilson Libr Bull* 1981, 56:256–60.
216. Hicks CR. *Fundamental concepts in the design of experiments,* 3rd ed. New York: Holt, Rinehart and Winston, 1982.
217. Hiltz SR, Turoff M. *The network nation: human communication via computer.* Reading, Mass.: Addison-Wesley, 1978.
218. Horrobin DF. Referees and research administration: barriers to scientific research? *Br Med J* 1974, 2:216–8.
219. Hutchins WJ. Machine translation and machine-aided translation. *J Doc* 1978, 34:119–59.
220. Huth EJ. *How to write and publish papers in the medical sciences.* Philadelphia: ISI Press, 1982.
221. Hyams J. Cells in motion. *Nature* 1982, 299:498.
222. Ingelfinger FJ. Medical literature: the campus without tumult. *Science* 1970, 169:831–7.

223. Ingelfinger FJ. Peer review in biomedical publication. *Am J Med* 1974, 56:686–92.
224. Ingelfinger JA et al. *Biostatistics in clinical medicine.* New York: Macmillan, 1983.
225. Institute for Scientific Information. *User's guide to ISI/BIOMED.* Philadelphia, no date, p. 2.
226. Instruction for contributors. *Science* 1968, 162(1):xv–xvi.
227. International Commission on Radiological Protection. *Report of the Task Group on Reference Man.* Oxford: Pergamon Press, 1975.
228. International Committee of Medical Journal Editors. Uniform requirements for manuscripts submitted to biomedical journals. *Ann Intern Med* 1982, 96:766–71; also *Br Med J* 1982, 284:1766–70.
229. *ISI Atlas of science: biochemistry and molecular biology, 1978/80.* Philadelphia: Institute for Scientific Information, 1981.
230. Jablonski S. *Illustrated dictionary of eponymic syndromes, and diseases, and their syndromes.* Philadelphia: Saunders, 1969.
231. Jackson BD. Authorship in the *Amoenitatis Academicae. J Bot* 1913, 51:101–3.
232. Jacobstein JM, Mersky RM. *Legal research illustrated.* Mineola, N.Y.: Foundation Press, 1980.
233. Jahoda G. *Information storage and retrieval systems for individual researchers.* New York: Wiley-Interscience, 1970.
234. Jarcho S. Some hoaxes in the medical literature. *Bull Hist Med* 1959, 33:342–6.
235. Johnson LG. Reprint requests. *Nature* 1973, 242:143.
236. Journals in microform. *Science* 1959, 129:30.
237. Juhasz S. Acceptance and rejection of manuscripts. *IEEE Trans Prof Commun* 1975, PC 18:177–84.
238. Kaplan J. *Lincoln Steffens: a biography.* New York: Simon and Schuster, 1974.
239. Kaplan N. The norms of citation behavior. *Am Doc* 1965, 16:179–84.
240. Katani M. Data generation. *In:* Rossmassler SA, Watson DG, eds. *Data handling for science and technology.* Amsterdam: North Holland, 1980, pp. 9–21.
241. Katzen MF. The changing appearance of research journals in science and technology: an analysis and a case study. *In:* Meadows AJ, *Development of science publishing in Europe.* Amsterdam: Elsevier Science Publishers, 1980.
242. Keele D. Cybernetic aspects of medical history. *Scott Med J* 1967, 12:256–9.
243. Kent A. Relevance predictability in information retrieval systems. *Methods Inf Med* 1967, 6:45–51.
244. Key JD, Roland CG. Reference accuracy in articles accepted for publication in the *Archives of Physical Medicine and Rehabilitation.* Arch Phys Med 1977, 58:136–7.
245. King A. Concerning conferences. *J Doc* 1961, 17:69–76.
246. King DW, McDonald DD, Roderer ND. *Scientific journals in the United States: their production, use and economics.* Stroudsburg, Pa.: Hutchinson Ross, 1981.
247. King DW, et al. *Statistical indicators of scientific and technical information: summary report.* Rockville, Md.; King Research Inc., 1976.
248. King LS, Roland CG. *Scientific writing.* Chicago: American Medical Association, 1968.
249. Kissman HM. Information retrieval in toxicology. *Annu Rev Pharmacol Toxicol* 1980, 20:285–305.
250. Kissmeyer-Nielsen F et al. Scandiatransplant: preliminary report of a kidney exchange program. *Transplant Proc* 1971, 3:1019–29.
251. Knox RA. Deeper problems for Darsee: Emory probe. *J Am Med Assoc* 1983, 249:2867, 2871–3, 2876.

252. Kobler J. *The reluctant surgeon; a biography of John Hunter.* New York: Doubleday, 1960, pp. 319–21.

253. Koch-Weser J. Inundation by requests for costly reprints. *N Engl J Med* 1976, 295:55.

254. Koestler A. *The call girls: a tragi-comedy.* New York: Random House, 1973.

255. Koestler A. *The case of the midwife toad.* New York: Random House, 1972.

256. Kohler RE. *From medicinal chemistry to biochemistry: the making of a biomedical discipline.* Cambridge: Cambridge University Press, 1982, p. 198.

257. Krebs HA. *Reminiscence and reflections.* Oxford: Clarendon Press, 1981.

258. Kremenetskaya AV, Vasil'eva EV. (Abstracted information). *Vestn Akad Nauk SSSR* 1952, 22(8/9):41–5. (Transl. and issued by the Technical Information Bureau Ministry of Supply, Great Britain, Jan 1954).

259. Kronick DA. Goodbye to farewells. *J Acad Librarianship* 1982, 8:132–6.

260. Kronick DA. *A history of scientific and technical periodicals,* 2nd ed. Metuchen, N.J.: Scarecrow Press, 1976.

261. Kronick DA. Literature citations, a clinico-pathological study, with presentation of three cases. *Bull Med Libr Assoc* 1958, 46:219–23.

262. Kronick DA. Nicolas de Blegny; medical journalist. *Bull Cleveland Med Lib* 1960, 7:47–56.

263. Kronick DA. On keeping up with new journals. *Bull Med Libr Assoc* 1982, 70:331–2.

264. Kuhn TS. *The essential tension; selected studies in scientific tradition and change.* Chicago: The University of Chicago Press, 1977.

265. Kuhn TS. The function of dogma in scientific research. *In:* Crombie AC, ed. *Scientific change.* New York: Basic Books, 1963, pp. 347–69.

266. Kuhn TS. *The structure of scientific revolutions.* Chicago: University of Chicago Press, 1962.

267. Lacy D. Publishing and the new technology. *In:* Segal E et al. *Books, libraries and electronics.* White Plains, N.Y.: Knowledge Industries, 1982, pp. 73–92.

268. Lancaster FW. *Evaluation of the MEDLARS demand search service.* Bethesda, Md.: National Library of Medicine, 1968.

269. Lancaster FW. *Libraries and librarians in an age of electronics.* Arlington Va.: Information Resources Press, 1982.

270. Lancaster FW. Vocabulary control in information retrieval systems. *Adv Librarianship* 1977, 7:1–41.

271. Lancaster FW, Smith LC. Science, scholarship and the communication of knowledge. *Libr Trends* 1978, 27:367–88.

272. Latour B, Woolgar S. *Laboratory life: the social construction of scientific facts.* Beverly Hills: Sage Publications, 1979.

273. Lederberg J. Digital communication and the conduct of science: the new literacy. *Proc IEEE* 1978, 66(11):1314–9.

274. Le Fanu J. Pasteur's notes tells a different story. *Med News* 1983, 7(11):9.

275. Liebesny F. Lost information: unpublished conference papers. *Proceedings of the International Conference on Scientific Information,* Washington, D.C. 16–21 Nov 1958. Washington, D.C.: National Academy of Sciences, 1959, pp. 475–9.

276. Line MB, Sandison A. "Obsolescence" and change in the use of literature with time. *J Doc* 1974, 30:283–350.

277. Line MB, Wood DN. The effect of a large-scale photocopying service on journal sales. *J Doc* 1975, 31:234–45.

278. Lipsett FR, Haycock AB. Preparation of personal bibliographies using a large computing facility. *J Chem Inf Comput Sci* 1982, 22:1–4.

279. Loeper M. L'evolution de la presse médicale. *Cahiers de la presse* 1936, 1:538.

280. Longmore JM. Keeping up to date. *Br Med J* 1979, 1:1547–8.

281. Lotka AJ. The frequency distribution of scientific productivity. *J Washington Acad Sci* 1926, 16:217–23.

282. Lowry OH, Rosebrough NJ, Farr AL, Randall RJ. Protein measurement with the Folin phenol reagent. *J Biol Chem* 1951, 193:265–75.

283. Lunt DA. Ghost authors. *Nature* 1974, 252:629.

284. Lynch BS, Chapman CF. *Writing for communication in science and medicine.* New York: Van Nostrand-Reinhold, 1980.

285. MacCurdy E, ed. *The notebooks of Leonardo da Vinci.* New York; George Brazilier, 1955.

286. McCutchen C. An evolved conspiracy. *New Sci* 1976, 70:225.

287. Machines break down language barrier. *New Sci* 1980, 85:836.

288. McMahon RE, Rawling DA, Weimann MR. An information filing system for oral surgeons. *Oral Surg* 1973, 35:160–7.

289. Magalini S. *Dictionary of medical syndromes,* 2nd ed. Philadelphia: Lippincott, 1981.

290. Mahy BWJ. Viral treatment. *Nature* 1982, 299:500.

291. Making journals fit for readers (editorial). *Nature* 1981, 293:341.

292. Manten AA. Future of the scientific journal. *Trends Biochem Sci* 1977, 2(12):269–70.

293. Manuila A. Medical terminology and medical dictionaries, an assessment of problems, needs and prospects. *In:* Council for International Organizations of Medical Sciences. *Medical terminology and lexicography.* Basel: Karger, 1966, pp. 27–40.

294. Manzer BM. *The Abstract Journal, 1790–1920.* Metuchen, N.J.: Scarecrow Press, 1977.

295. Margolis J. Citation indexing and evaluation of scientific papers. *Science* 1967, 155:1213–9.

296. Markey K. Interindexer consistency tests. Libr Inf Scl Res 1984, 6:155–77.

297. May KO. Abuses of citation indexing. *Science* 1967, 156:890–2.

298. May KO. Growth and quality of the mathematical literature. *Isis* 1968, 59:363–71.

299. Meadows AJ. *Communication in science.* London: Butterworth, 1974.

300. Medawar PB. *The art of the soluble.* London: Methuen and Co., 1967.

301. Medawar PB. Is the scientific paper fraudulent? *Saturday Rev* 1964, Aug 1:42–3.

302. Medawar PB. Science and literature. *Pespect Biol Med* 1969, 12:529–46.

303. Meiss HR, Jaeger DA. Comp. *Information to authors, 1980–81: editorial guidelines reproduced from 246 medical journals.* Baltimore: Urban and Schwarzenberg, 1980.

304. Meredith P. *Instruments of communication: an essay on scientific writing.* Oxford: Pergamon Press, 1966.

305. Merton RK. The ambivalence of scientists. *Bull Johns Hopkins Hosp* 1963, 112:77–97.

306. Merton RK. The ambivalence of scientists. In: *The sociology of science.* Chicago; University of Chicago Press, 1973, pp. 383–412.

307. Merton RK. Foreword to: Garfield E, *Citation indexing.* New York: Wiley, 1979.

308. Merton RK. The Matthew effect in science. *Science* 1968, 159:56–63.

309. Merton RK. The normative structure of science. In: *The sociology of science.* Chicago: University of Chicago Press, 1973, pp. 267–85.

310. Merton RK. *On the shoulders of giants.* New York: Harcourt, Brace and World, 1965.

311. Merton RK. *The sociology of science: an episodic memoir.* Carbondale: Southern Illinois University Press, 1977.

312. Morris RT, Holtum EA, Curry DS. Being there: the effect of the user's presence on Medline search results. *Bull Med Libr Assoc* 1982, 70:298–304.

313. Mosteller F. Evaluation: requirements for scientific proof. In: Warren KS, ed. *Coping with the biomedical literature.* New York: Praeger, 1981, pp. 103–21.

314. Mosteller F. Problems of omission in communication. *Clin Pharmacol Ther* 1979, 25:761–4.

315. Murphy EA. *Biostatistics in medicine.* Baltimore: Johns Hopkins University Press, 1982.

316. Narin F, Carpenter MP. National publication and citation comparisons. *J Am Soc Inf Sci* 1975, 26:80–93.

317. Nation EF. William Osler on penis captivus and other urologic topics. *Urology* 1973, 2:468–70.

318. National Academy of Sciences. Committee on Research in the Life Sciences. *The life sciences.* Washington, D.C.: National Academy of Sciences, 1970.

319. National Academy of Sciences. Committee on Scientific and Technical Communication. *Scientific and technical communications.* Washington, D.C.: National Academy of Sciences, 1969, p. 125.

320. National Center for Health Statistics. *International classification of diseases,* 9th revision. Washington, D.C.: Public Health Service, 1980.

321. National Library of Medicine. *Medical Subject Headings 1984. Index Medicus* 1984, 25 (1 pt. 2).

322. National Library of Medicine. Categories and subcategories. In: *Medical subject headings 1983.* Washington, D.C. *Index Medicus* 1983, 24 (1 pt. 2).

323. National Library of Medicine. *Classification, 4th ed. 1978, revised 1981.* Washington, D.C., 1981.

324. National Library of Medicine. *Program and services FY 1981.* Bethesda, Md., 1982.

325. Neavill GB. Electronic publishing, libraries and the survival of information. *Libr Res Tech Ser* 1984, 28:76–89.

326. Nelkin D. Intellectual property: the control of scientific information. *Science* 1982, 216:704–8.

327. New journals June 1980 to May 1981. *Nature* 1982, 299:491–2.

328. New journals' review. *Nature* 1981, 293:341–68.

329. Newmark P. A layman's view of medical translation. *Br Med J* 1979, 2:1405–7.

330. Nimo J. The ideology of scientific evaluation. *Trends Biochem Sci* 1981, 6(7):vii–ix.

331. Norris C. MeSH, the subject heading approach. *ASLIB Proc* 1981, 33:153–9.

332. O'Connor M. *Editing scientific books and journals.* Tunbridge Wells: Ritman Medical, 1978.

333. O'Connor M. *The scientist as editor.* New York: Wiley, 1979.

334. O'Connor M. Standardization of bibliographical reference systems. *Br Med J* 1978, 1:31–2.

335. O'Connor M, Woodford FP. *Writing scientific papers in English.* New York, Excerpta Medica, 1976.

336. Office of Technology Assessment. *Medlars and health information policy; a technical memorandum.* Washington, D.C. 1982.

337. Ogburn WF, Thomas D. Are inventions inevitable? A note on social evolution. *Political Sci Q* 1922, 37:83–98.

338. Oliver S. The current ferment in publishing. *Nature* 1981, 293:354.

339. Online databases grow in number and revenues. *Bull Am Soc Inf Sci* 1983, 8(8):6.

340. *Online Newsletter.* Nov 1982, p. 10.

341. Onuigbo WIB. Printer's devil and reprint requests. *J Am Soc Inf Sci* 1982, 33:58–9.

342. Orr R. The metabolism of scientific information: a preliminary report. *Am Doc* 1961, 12:15–9.

343. Orr RH, Leeds AA. Biomedical literature: volume, growth and other characteristics. *Fed Proc* 1964; 23:1310–31.

344. Ortega y Gasset J. *Revolt of the masses.* New York: Norton, 1932. (Cited in Cole and Cole, ref. 80, p. 217.)

345. Osler W. *Bibliotheca Osleriana.* Montreal: McGill-Queens University Press, 1969, p. 117.

346. Osler W. *The principles and practice of medicine.* New York: Appleton, 1892.

347. Osler W. The student life. In: *Acquanimitas, with other addresses.* London: H. K. Lewis, 1904, pp. 395–423.

348. Over R. The durability of scientific reputation. *J Hist Behav Sci* 1982, 18:53–61.

349. Padin ME. At ease with Boolean operators in online searching. *Online* 1982, 6(Apr):82–5.

350. Palmer RC. *Online reference and information retrieval.* Littleton, Colo.: Libraries Unlimited, 1983.

351. Papadakis EP. What drops off the end. *Am Libr* 1982, 13:575–6.

352. Paris J. Basics of videodisc and optical disk technology. *J Am Soc Inf Sci* 1983, 34(6):408–13.

353. Passman S. *Scientific and technological communication.* London: Pergamon, 1969.

354. Pasteur L. (address, 11 Sept 1872) *C R des travaux du Congres viticale et sericiole de Lyon,* 9–14 Sept. 1872, p. 49.

355. Pasteur L. Melanges scientifiques et litteraires. *Oeuvres,* VII, Paris, Masson, 1930, pp. 129–32.

356. Pearson LD. A cross-indexing system for a personal information file. *J Am Vet Med Assoc* 1970, 157:344–8.

357. Peters DP, Ceci CJ. Peer review practices of psychological journals: the fate of published articles submitted again. *Behav Brain Sci* 1982, 5:185–95.

358. Peters DP, Ceci CJ. Peer review research: objectives and obligations. *Behav Brain Sci* 1982, 5:246–55.

359. Phelps RH, Herling JP. Alternatives to the scientific periodical. *Unesco Bull Libr* 1960, 14(2):2–12.

360. Pocker BB. "User publishing" a new concept in electronic publishing. *Pub Weekly,* 1982, 221(10):38–9.

361. Porter AL et al. The role of the dissertation in scientific careers. *Am Sci* 1982, 70:475–81.

362. Powell RH, ed. *Handbook and tables in science and technology.* Phoenix, Ariz.: Oryx Press, 1979.

363. Pownall JF. *Organized publication.* London: Elliott Stock, 1926.

364. Poyer RK. Inaccurate references in significant journals of sciences. *Bull Med Libr Assoc* 1979, 67:396–8.

365. Poyer RK. *Science Citation Index's* coverage of the preclinical science literature. *J Am Soc Inf Sci* 1982, 33:317–20.
366. Poyer RK. Time lag in four indexing services. *Spec Libr* 1982, 73(2):142–6.
367. President's Commission on Foreign Language and International Studies. *Strength through wisdom: a critique of U.S. capability.* Washington, D.C., 1979.
368. Price DJ. The citation cycle. In: Griffith BC, ed. *Key papers in information science.* White Plains: Knowledge Industries Publications, 1980, pp. 195–210.
369. Price DJ. A general theory of bibliometric and other cumulative advantage processes. *J Am Soc Inf Sci* 1976, 27:292–306.
370. Price DJ. Is technology historically independent of science? *Technol Cult* 1965, 6:553–68.
371. Price DJ. Networks of scientific papers. *Science* 1965, 149:510–5.
372. Price DJ. *Science since Babylon.* New Haven: Yale University Press, 1961.
373. Price DJ, Beaver DD. Collaboration in an invisible college. *Am Psychol* 1966, 21:1011–8.
374. Publication and classification of scientific knowledge. *Nature* 1947, 160:649–50.
375. Publish and be damned a second time. *Nature* 1971, 233:294.
376. Ratnoff OD. How to read a paper. *In:* Warren KS, ed. *Coping with the biomedical literature.* New York: Praeger, 1981, pp. 95–101.
377. Ravetz JR. *Scientific knowledge and its social problems.* Oxford: Clarendon Press, 1971.
378. Reeder RC, ed. *Sourcebook of medical communication.* St. Louis: Mosby, 1981.
379. Relman A. Journals. *In:* Warren KS, ed. *Coping with the biomedical literature.* New York: Praeger, 1981, pp. 67–78.
380. Relman AS. Lessons from the Darsee affair. *N Engl J Med* 1983, 308:1415–7.
381. Rescher R. *Introduction to logic.* New York: St. Martin's Press, 1964.
382. Riegelman RK. *Studying a study and testing a test: how to read the medical literature.* Boston: Little Brown, 1981.
383. Rohlfs H. Eine literarische legende. *Dtsch Arch Ges Med* 1880, 3:270–2.
384. Rosenbaum RA et al. Academic freedom and the classified information system. *Science* 1983, 219:257–9.
385. Rosenthal R. Reliability and bias in peer-review practices. *Behav Brain Sci* 1982, 5:235–6.
386. Rosner BA. *Fundamentals of biostatistics.* Boston: Duxbury Press, 1982.
387. Rowland JFB. Synopsic journals as seen by their authors. *J Doc* 1981, 37:69–76.
388. Rowley JC. The conference literature. *In:* Zamara G, Adamson MC, eds. *Conference literature: its role in the dissemination of information.* Marlton, N.J.: Learned Information Inc., 1981. pp. 11–21.
389. Sackett DL. Bias in analytical research. *J Chronic Dis* 1979, 32:51–63.
390. Sackett DL. Evaluation: requirements for clinical application. In: Warren KS, ed. *Coping with the biomedical literature.* New York: Praeger, 1981, pp. 123–57.
391. St. James-Robert I. Cheating in science. *New Sci* 1976, 72:466–9.
392. Saracevic T. Relevance: a review of the literature and framework for thinking on the notion in information in science. *Adv Libr* 1976, 6:79–138.
393. Scheckler WE. A realistic journal reading plan. *J Am Med Assoc* 1982, 248:1987–8.
394. Schor S, Karten I. Statistical evaluation of medical journal manuscripts. *J Am Med Assoc* 1966, 195:145–50.

395. Schutt DC. Teaching reference file for family practice residencies. *Milit Med* 1981, 146:336–8.
396. *Science Citation Index Annual Guide.* Philadelphia: Institute for Scientific Information, 1981.
397. *Séances publique tenue par la Faculté de Medécine.* Paris, 1789.
398. Seiler LH, Robin J. The electronic journal. *Society* 1981, 18:76–83.
399. Sewell W. *Guide to drug information.* Hamilton, Ill.: Drug Intelligence, 1976.
400. Sheplan L. An inexpensive simple method of data retrieval. *S Med J* 1969, 62:418–20.
401. Shera JH. On keeping up with keeping up: recent trends in document storage and retrieval. *UNESCO Bull Lib* 1962, 16(Mar):64–72.
402. Shervis LJ, Shenefelt RD. Poor access to Apanteles species literature through titles, abstracts and automatically extracted species varies as keywords. *J Wash Acad Sci* 1973, 63:22–5.
403. Shifrine M, McMartin DA. How to file information—and find it. *J Am Vet Med Assoc* 1961, 138:613–5.
404. Shils E. The confidentiality and anonymity of assessment. *Minerva* 1975, 13:135–51.
405. Siesjo BK. A new journal and a new society—why? *J Cereb Blood Flow Metab* 1981, 1:3–4.
406. Sigerist H. Nationalism and internationalism in medicine. *Bull Hist Med* 1947, 21:5–16.
407. Silver S. Ethical questions in the peer review system. *ASM News* 1980, 46:302–6.
408. Simon HA. *Models of thought.* New Haven: Yale University Press, 1979.
409. Singer DW. Some plague tracts (14th and 15th centuries). *Proc R Soc Med Hist Soc* 1916, 9:159–212.
410. Singer K. Where did I see that article? *J Am Med Assoc* 1979, 241:1492–3.
411. Slocum HE. Personal medical reference files for family physicians. *J Fam Prac* 1977, 5:593–5.
412. Small H. Co-citation in the scientific literature: a new measure of the relationship between two documents. *J Am Soc Inf Sci* 1973, 24:265–9.
413. Small H, Griffith BC. The structure of scientific literature. *Sci Stud* 1974, 4:17–40, 339–65.
414. Smith L. Citation analysis. *Libr Trends* 1981, 30:83–106.
415. Soergel D. *Indexing languages and thesauri: construction and maintenance.* Los Angeles: Melville Publishing Co., 1974.
416. Sorell W. Facets of creativity. *MD* 1979, 23(12):9–11.
417. Sprat T. *The history of the Royal Society of London.* London: Martyn, 1667, pp. 112–3.
418. Stanford denies cover-up of research fraud. *NY Times,* 23 Aug 1981, p. 31.
419. Stankus T. Collection development: journals for biochemists. *Spec Collections* 1982, 2:51–69.
420. Starr P. The electronic reader. *Daedalus* 1983, 112:143–56.
421. Stibic V. *Personal documentation for professionals.* Amsterdam: North-Holland, 1980.
422. Stoldal PM, Gordon DB. Uniformity of references. *Science* 1974, 186:1158–9.
423. Strunk W, White EB. *The elements of style,* 3rd ed. New York: Macmillan, 1979.
424. Stumpf WE. Peer review. *Science* 1980, 207:822–3.

206   *References*

425. Subramanyan K. *Scientific and technical information resources.* New York: Dekker, 1981.
426. Swanson DR. Information retrieval as a trial-and-error process. *Libr Q* 1977, 47:128–48.
427. Swanson DR. Review of: Lancaster FW. ed. *The role of the library in an electronic society.* Urbana: University of Illinois Graduate School of Library Science, 1980, *Libr Q* 1981, 51:316–8.
428. Tagliacozzo R. Self-citations in scientific literature. *J Doc* 1977, 33:251–5.
429. Taylor FK. Penis captivus—did it occur? *Br Med J* 1979, 2:977–8.
430. Tedd LA. Software for microcomputers in libraries and information units. *Electron Lib* 1983, 1:31–48.
431. Teigen PM, Bensley EH. An Egerton Y. Davis checklist. *Osler Libr Newslett* 1981, 38:1–5.
432. Ten is limit on authors cited in Index Medicus and Medline. *Natl Libr Med News* 1983, 38(12):2.
433. *Thesaurus of engineering and scientific terms.* New York: Engineers Joint Council, 1967.
434. Thomas GJ. Editing a basic science journal. *IEEE Trans Prof Commun* 1975, PC18:186–9.
435. Thompson DW. *On growth and form,* 2nd ed. Cambridge: Cambridge University Press, 1942.
436. Thorwald J. *The century of the detective.* New York: Harcourt, Brace, 1965.
437. Three publishers drop out of ADONIS, project's future uncertain. *Adv Tech Libr* 1983, 12(4):1.
438. Toong HD, Gupta A. Personal computers. *Sci Am* 1982, 247(12):87–105.
439. Trachtman LE. The public understanding of science effort: a critique. *Sci Technol* 1981, 36:10–5.
440. Turner RS. Helmholtz. In: *Dictionary of scientific biography.* New York, Scribners, VI:241–53.
441. Turoff M, Hiltz SR. The electronic journal: a progress report. *J Am Soc Inf Soc* 1982, 33:195–202.
442. Tweney RD, Doherty ME, Mynatt CR, eds. *On scientific thinking.* New York: Columbia University Press, 1981.
443. *Ulrich's international periodical directory,* 22nd ed. New York: Bowker, 1983.
444. Ultmann JE. More on duplicate abstracts and filing. *Ann Intern Med* 1968, 68:711–2.
445. U.S. Office of Naval Records. *Thesaurus of engineering and scientific terms.* Alexandria, Va., 1967.
446. University of Chicago Press. *The Chicago manual of style,* 13th ed. Chicago: University of Chicago Press, 1982.
447. Van den Daele W. The social construction of science. In: Mendelsohn E et al., eds. *The social production of scientific knowledge.* Dordrecht: D. Reidel, 1977, pp. 27–54.
448. Venulet J, Blattner R, Von Bulow J, Berneker GC. How good are articles on adverse drug reactions? *Br Med J* 1982, 284:252–4.
449. Vickery BC. Statistics of scientific and technical articles. *J Doc* 1968, 24:192–6.
450. Virtanen R. Claude Bernard and the history of ideas. In: Grande F, Visscher MB, eds. *Claude Bernard and experimental medicine.* Cambridge, Mass.: Schenkman, 1967, pp. 9–28.

451. Visscher MB. Copyright and other impediments to scientific communication. In: Day SM, ed. *Communication of scientific information.* Basel: Karger, 1975, pp. 118–28.
452. Wade N. The all-too-human face of the Nobel prize. *New Sci* 1981, 92:251–5.
453. Wade N. *The Nobel duel.* Garden City, N.Y.: Anchor Press/Doubleday, 1981.
454. Waksman BH. Information overload in immunology: possible solutions to the problem of excessive publication. *J Immunol* 1980, 124:1009–15.
455. Walls E. Term relationships. *J Am Soc Inf Sci* 1975, 26:71–9.
456. Warren KS, ed. *Coping with the biomedical literature: a primer for the scientist and the clinician.* New York: Praeger, 1981.
457. Warren KS. Selective aspects of the biomedical literature. In: *Coping with the biomedical literature.* New York: Praeger, 1981, pp. 17–30.
458. Watson JD, Crick FHC. A structure for deoxyribose nucleic acid. *Nature* 1953, 171:737–8.
459. Webb EC. Communication in biochemistry. *Nature* 1970, 225:132–5.
460. Weiss P. Knowledge, a growth process. *Science* 1960, 131:1716–9.
461. Wells HG. *World brain.* New York: Doubleday, 1938.
462. Werner G. The use of on-line bibliographic retrieval services in health science libraries in the United States and Canada. *Bull Med Libr Assoc* 1979, 67:1–14.
463. White HD, Griffith BC. Author co-citation: literature measure of intellectual structure. *J Am Soc Inf Sci* 1981, 32:163–71.
464. White HS. Publishers, libraries, and cost of journal subscriptions in times of funding retrenchments. *Libr Q* 1976, 46:359–77.
465. Williams ME, ed. *Computer-readable databases.* Washington, D.C.: American Society for Information Science, 1979.
466. Wilson EB. *An introduction to scientific research.* New York: McGraw-Hill, 1952.
467. Wilson JD. Peer review and publication. *J Clin Invest* 1978, 61:1697–701.
468. Wintrobe MM. *Clinical hematology,* 8th ed. Philadelphia: Lea & Febiger, 1981.
469. Wither G. The lover's resolution. In: *The Oxford book of English verses, 1250–1918.* Quillen-Couch A, ed. New York: Oxford University Press, 1939, pp. 262–3.
470. Wizola SJ. How to choose a portable. *Byte* 1983, 8(9):34–47.
471. Wood DN. The foreign language problem facing scientist and technologists in the United Kingdom. *J Doc* 1966, 23:117–30.
472. Wood JL, Flanagan C, Kennedy HE. Overlap in the lists of journals monitored by BIOSIS, CAS, and EI. *J Am Soc Inf Sci* 1972, 23:36–8.
473. Wood JL, Flanagan C, Kennedy HE. Overlap among the journal articles selected for coverage by BIOSIS, CAS and EI. *J Am Soc Inf Sci* 1973, 24:25–8.
474. Woodward AM. Review literature: characteristics, source and output in 1972. *ASLIB Proc* 1974, 26:367–76.
475. Ziman J. Information, communication, knowledge. *Nature* 1969, 224:318–24.
476. Ziman J. *Public knowledge.* Cambridge: Cambridge University Press, 1968.
477. Ziman JM. Teaching scientists to find information. In: *Puzzles, problems and enigmas.* Cambridge: Cambridge University Press, 1981, pp. 315–8.
478. Zimmerman LM. The beginnings of surgery and the "Edwin Smith papyrus." *J Int Coll Surg* 1957, 27(1, Sec II):14–21.
479. Zinder N. Review of *Betrayers of the Truth. Science 83* 1983, 4:94–5.
480. Zinsser H. *Rats, lice and history.* New York: Blue Ribbon Books, 1938.
481. Zinsser W. *On writing well: an informal guide to writing non-fiction,* 2nd ed. New York: Harper and Row, 1980.

482. Zipf GK. *Human behavior and the principle of least effort.* Cambridge, Mass.: Addison Wesley, 1949.
483. Zuckerman H, Merton RK. Age, aging and age structure in science. In: Merton RK. *The sociology of science.* Chicago, The University of Chicago Press, 1973, pp. 497–559.
484. Zuckerman H, Merton RK. Patterns of evaluation in science. *Minerva* 1971, 9:66–100.

Index